101
AMERICAN GEO-SITES
you've gotta see

101 AMERICAN GEO-SITES
you've gotta see

Albert Binkley Dickas

2012
Mountain Press Publishing Company
Missoula, Montana

© 2012 by Albert Binkley Dickas
First Printing, April 2012
All rights reserved

Fifth Printing, December 2017

Photos © 2012 by Albert Binkley Dickas unless otherwise credited

Cover photo: Boulder Flatirons, © by John Karachewski
Illustrations constructed by Mountain Press Publishing based on original drafts by the author

Library of Congress Cataloging-in-Publication Data

Dickas, Albert B.
 101 American geo-sites you've gotta see / Albert Binkley Dickas.
       p. cm.
 Includes bibliographical references and index.
 ISBN 978-0-87842-587-7 (pbk. : alk. paper)
 1. Landforms—United States—Guidebooks. 2. United States—Guidebooks.
 I. Title.
 GB400.6.D53 2012
 557.3—dc23
                                            2012000030

PRINTED IN HONG KONG

P.O. Box 2399 • Missoula, MT 59806 • 406-728-1900
800-234-5308 • info@mtnpress.com
www.mountain-press.com

I dedicate this volume to my grandchildren:
Emma Margaret
Gabrielle Nicole
James Douglas
William Elias
Kathryn Adrienne
Thomas Christian
Jessica Louise
Braden Patrick
May they grow to appreciate the wonders of nature.

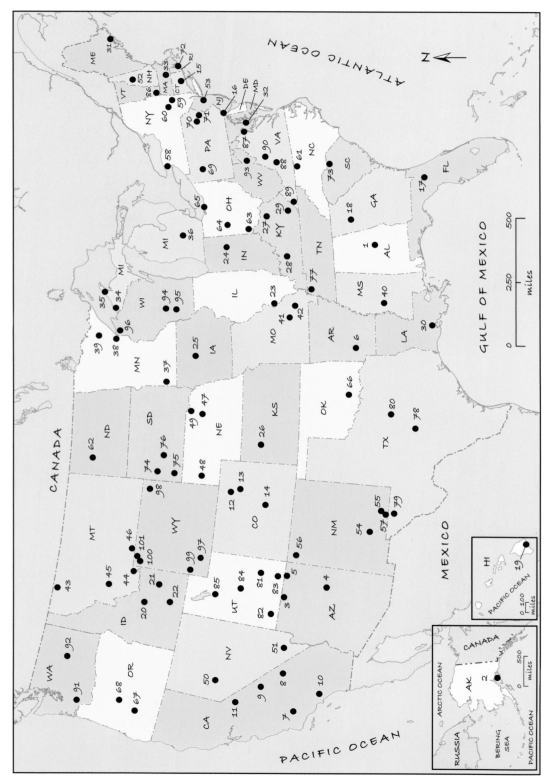

Locations of geo-sites in the book. The numbers correspond to specific sites.

Contents

Foreword ix

Acknowledgments xi

Introduction 1

Relative Age Dating 1

Absolute Age Dating 4

Plate Tectonics Theory 5

A Short History of the Earth 11

Hadean Eon: 4,600 to 4,000 Million Years Ago 11
Archean Eon: 4,000 to 2,500 Million Years Ago 12
Proterozoic Eon: 2,500 to 542 Million Years Ago 12
Cambrian Period: 542 to 488 Million Years Ago 12
Ordovician Period: 488 to 443 Million Years Ago 13
Silurian Period: 443 to 416 Million Years Ago 13
Devonian Period: 416 to 359 Million Years Ago 14
Carboniferous Period: 359 to 299 Million Years Ago 14
Permian Period: 299 to 251 Million Years Ago 15
Triassic Period: 251 to 199 Million Years Ago 15
Jurassic Period: 199 to 145 Million Years Ago 15
Cretaceous Period: 145 to 65 Million Years Ago 16
Paleogene Period: 65 to 23 Million Years Ago 16
Neogene Period: 23 to 2.6 Million Years Ago 17
Quaternary Period: 2.6 Million Years Ago to Today 17

GEO-SITES

1. Wetumpka Crater, Alabama 20
2. Exit Glacier, Alaska 22
3. Antelope Canyon, Arizona 24
4. Meteor Crater, Arizona 26
5. Monument Valley, Arizona 28
6. Prairie Creek Pipe, Arkansas 30
7. Wallace Creek, California 32
8. Racetrack Playa, California 34
9. Devils Postpile, California 36
10. Rancho La Brea, California 38
11. El Capitan, California 40
12. Boulder Flatirons, Colorado 42
13. Interstate 70 Roadcut, Colorado 44
14. Florissant Fossil Beds, Colorado 46
15. Dinosaur Trackway, Connecticut 48
16. Wilmington Blue Rocks, Delaware 50
17. Devil's Millhopper, Florida 52
18. Stone Mountain, Georgia 54
19. Kilauea Volcano, Hawaii 56
20. Borah Peak, Idaho 58
21. Menan Buttes, Idaho 60
22. Great Rift, Idaho 62
23. Valmeyer Anticline, Illinois 64
24. Hanging Rock Klint, Indiana 66
25. Fort Dodge Gypsum, Iowa 68
26. Monument Rocks, Kansas 70
27. Ohio Black Shale, Kentucky 72
28. Mammoth Cave, Kentucky 74
29. Four Corners Roadcut, Kentucky 76
30. Avery Island, Louisiana 78
31. Schoodic Point, Maine 80
32. Calvert Cliffs, Maryland 82
33. Purgatory Chasm, Massachusetts 84
34. Nonesuch Potholes, Michigan 86
35. Quincy Mine, Michigan 88
36. Grand River Ledges, Michigan 90
37. Sioux Quartzite, Minnesota 92
38. Thomson Dikes, Minnesota 94
39. Soudan Mine, Minnesota 96
40. Petrified Forest, Mississippi 98
41. Elephant Rocks, Missouri 100
42. Grassy Mountain Nonconformity, Missouri 102
43. Chief Mountain, Montana 104

44. Madison Slide, Montana   106
45. Butte Pluton, Montana   108
46. Quad Creek Quartzite, Montana   110
47. Ashfall Fossil Beds, Nebraska   112
48. Scotts Bluff, Nebraska   114
49. Crow Creek Marlstone, Nebraska   116
50. Sand Mountain, Nevada   118
51. Great Unconformity, Nevada   120
52. Flume Gorge, New Hampshire   122
53. Palisades Sill, New Jersey   124
54. White Sands, New Mexico   126
55. Carlsbad Caverns, New Mexico   128
56. Ship Rock, New Mexico   130
57. State Line Outcrop, New Mexico   132
58. American Falls, New York   134
59. Taconic Unconformity, New York   136
60. Gilboa Forest, New York   138
61. Pilot Mountain, North Carolina   140
62. South Killdeer Mountain, North Dakota   142
63. Hueston Woods, Ohio   144
64. Big Rock, Ohio   146
65. Kelleys Island, Ohio   148
66. Interstate 35 Roadcut, Oklahoma   150
67. Mount Mazama, Oregon   152
68. Lava River Cave, Oregon   154
69. Drake's Folly, Pennsylvania   156
70. Hickory Run, Pennsylvania   158
71. Delaware Water Gap, Pennsylvania   160
72. Beavertail Point, Rhode Island   162
73. Crowburg Basin, South Carolina   164
74. Mount Rushmore, South Dakota   166
75. Mammoth Site, South Dakota   168
76. Pinnacles Overlook, South Dakota   170
77. Reelfoot Scarp, Tennessee   172
78. Enchanted Rock, Texas   174
79. Capitan Reef, Texas   176
80. Paluxy River Tracks, Texas   178
81. Upheaval Dome, Utah   180
82. Checkerboard Mesa, Utah   182
83. San Juan Goosenecks, Utah   184
84. Salina Canyon Unconformity, Utah   186
85. Bingham Stock, Utah   188
86. Whipstock Hill, Vermont   190
87. Great Falls, Virginia   192
88. Natural Bridge, Virginia   194
89. Millbrig Ashfall, Virginia   196
90. Catoctin Greenstone, Virginia   198
91. Mount St. Helens, Washington   200
92. Dry Falls, Washington   202
93. Seneca Rocks, West Virginia   204
94. Roche-A-Cri Mound, Wisconsin   206
95. Van Hise Rock, Wisconsin   208
96. Amnicon Falls, Wisconsin   210
97. Green River, Wyoming   212
98. Devils Tower, Wyoming   214
99. Fossil Butte, Wyoming   216
100. Steamboat Geyser, Wyoming   218
101. Specimen Ridge, Wyoming   220

Glossary   223
References   229
Index   243

Foreword

Several years ago, I was traveling in the central part of Italy with friends near Gubbio, a place where a father-and-son team of scientists had begun to develop a theory that ultimately solved—albeit with controversy—one of the long-standing mysteries of evolutionary history. Some three decades earlier, Luis and Walter Alvarez had found a 1-centimeter-thick layer of red clay embedded in a limestone hillside. They recognized its potential importance and used the results of laboratory analysis to suggest that the abrupt extinction of the dinosaurs had an extraterrestrial cause.

Excited, I sought out a tourist office and asked for directions to the exact spot of their discovery, while my traveling companions waited not altogether patiently. Guided by a poorly drawn map and assurance that the "you can't miss it" exposure was but a few kilometers outside the medieval hilltown that has become a popular site for American tourists, we set off—my companions quizzically, and me with mounting anticipation. Sure enough, just beside the two-lane road with aptly described devil-take-the-hindmost Italian drivers whizzing by, there it was, identified by a beat-up sign.

We stopped; I scrambled, pointed, and posed proudly; and my companions took the requisite pictures. Back in the car, full of excitement since it is not every day that I have the opportunity to visit an iconic site of scientific importance, I told the story behind the clay. That thin layer of rock contains soot and also iridium, an element more commonly found in meteorites than on Earth. That combination was, to the scientists, telling evidence of a cataclysmic event resulting in a worldwide dust storm that blotted out the sun, changed the environment, and caused the extinction of most of the fauna then inhabiting Earth. The Alvarezes had found a plausible explanation for the seemingly overnight demise of the dinosaurs 65 million years ago—some burned out by massive fires, others suffocated by the oxygen-deprived atmosphere, all quickly deprived of the very world to which they had become accustomed.

Tens of thousands of people pass this outcrop annually, yet few know the background. For several years, this was the most famous rock exposure in the world. The history of life was rewritten by what the Alvarezes learned from the rocks collected there. My companions were well-educated, widely traveled people who had heard of the discovery but could see little in the outcrop to merit a second glance. Only after they heard the full story the clay layer held did the event come alive.

In part, this book is a result of that experience. I've spent six decades in the study of geology, a discipline that is filled with controversy, drama, and the continuing search for adequate explanations of the processes that affect our planet. Geology gives birth to lifelong investigation, if you have the curiosity, the know-how, and the time. Since at least AD 79, when Pliny the Elder left a record of a volcanic eruption that survived the destruction of Pompeii (although he didn't), geologists have investigated geo-sites such as the one at Gubbio, looking for the catalog of geologic events preserved within their rock foundations.

James Hutton, often called the father of geology, added a chapter when he stood on the rock-bound shoreline at Siccar Point, on the east coast of Scotland, with two dubious companions in 1788. Doubt changed to belief when he explained that the disturbed rock at their feet bespoke an Earth far older than the six millennia then accepted by society. Two centuries and two decades later, during my second visit to Siccar Point, my traveling companion, having unexpectedly been pulled through barbed-wire fences, across pastures punctuated with cow patties, and over rain-soaked rocks, shook his head and mumbled, "So?" I relished how his face changed when I recounted the story. "So?" changed to "Wow." Similar thoughts come to my mind when I recall visiting the fields of Wales where William Smith drew "the map that changed the world," or gazing upon the ice field that holds the grave of Alfred Wegener, who conceived the basic concepts of the theory of plate tectonics.

Of course, not everyone has the time, means, or temperament to travel to out-of-the-way places across continents. No problem, for such sites abound in the United States, if one knows where the right rocks are and how they should be "read." In fact, not 100 miles from where I now live is a hillside not at all unlike the one near Gubbio. Yes, this rail bed site is a bit more remote, and the travelers who pass by are on foot or in a locomotive, but there are similarities—a layer of clay distinguished by its reddish tone and imbedded in massive layers of limestone. It, too, is little to look at, but the story it holds is no less compelling. Rather than meteoritic impact, this site harbors evidence of perhaps the greatest volcanic eruption to have occurred anywhere on Earth in the last 500 million years.

This book presents the best of the geo-sites that the fifty states offer. Not every one is visually rich, but without exception each is a memorable source of information about the changes Earth continues to experience based on the oldest evidence we have: rocks. Despite the consensus of scientific interpretation associated with many of these stories, none of them is static in presentation. Different points of view, competing hypotheses, and the information afforded by new technology imbue these tales with a dynamism that is not likely to die. Indeed, Gubbio is a case in point. The Alvarezes' theory of meteoric impact is yet to be universally accepted, but in 2010 it received a confidence vote from a panel of esteemed scientists.

Referring to the beginning of the twentieth century, famed naturalist Edward O. Wilson wrote, "The highest mountains were still unclimbed, the ocean depths never visited, the vast wildernesses stretched across equatorial continents. Now we have all but finished mapping the physical world." Not entirely true. Even in the twenty-first century there are numerous locales that call for further study.

This guide illustrates and explains many that have stood the test of complete review, and others where mystery trumps history. It is up to you, the curious traveler, to decide which is which. As Proust so famously declared, "The real voyage of discovery consists not in seeking new landscapes but in having new eyes."

As for me, I'm headed back to Gubbio this summer.

Albert Binkley Dickas
Brush Mountain, Virginia
January, 2012

Acknowledgments

The writing of a field guide such as this, highlighting locales of unusual geology in all fifty of the United States cannot be accomplished without the help and cooperation of many individuals.

Richard Ojakangas, now retired from the University of Minnesota-Duluth, used his blue (actually, in his case it was red) pen wisely to keep my early efforts on an even keel. Unfortunately, his long-term writing commitments precluded our hoped-for collaboration on this project. Steven Uchytil, senior geological advisor with the Hess Corporation, reviewed the text with the exactitude of the corporate mind. His suggestions and recommendations brought a needed sense of clarity to the final product.

From day one of our association, my editor at Mountain Press, James Lainsbury, has been generous in encouragement, excessive in patience, and thorough in editing. It was his difficult task to inform me my first draft was much too long. I still marvel at his ability to create a higher degree of consistency in my use of dates, facts, and concepts. The pen and ink drawings used to great effect in this book are the work of Patsy Faires of Kernersville, NC. Her insightful interpretations are proof that black-and-white art can transcend the allure of color photography.

One individual deserves special recognition: Rachael M. Garrity, principal of Penworthy LLC, a published author in her own right, and a valued and cherished friend, tirelessly responded to my frequent requests for assistance in matters of grammar, usage, and syntax. She remains the very soul of tolerance, and her understanding and guidance have been indispensible during those many occasions when my flights of compositional fancy made little sense. My debt to her can never be repaid, but I shall try.

It is my good fortune to live within an easy drive of the library of the Virginia Polytechnic Institute and State University (Virginia Tech). Its staff and catalog of holdings have been an invaluable resource when I felt the need for geologic information, whether arcane or topical.

From conception to completion, I have been involved with this project for five years. During that time, I have driven the equivalent of the distance from shore to shore of the contiguous states several times and logged many hours in the air reaching distant geo-sites. I was familiar with the geology of many of them, but the what, why, and when of others were questions I needed to answer. In all cases, upon arrival at a locale I immediately sought out the principal experts, eager to know their thoughts on the local geology. In the ensuing months, I commonly followed up by telephone or email with additional questions and needs. Without exception, the enthusiasm and cooperation I experienced were the lights that shone during those days when the tunnel seemed darkest. Their count is in the hundreds, and space does not allow for individual recognition. To each and every one—at universities and colleges; federal, state, and community parks; and geological surveys from the state to the federal level—I say a heartfelt "thank you."

Three final statements need to be made: All photographs within this field guide without credits were taken by me. Any mistakes found herein are mine, and mine alone. And please be careful when visiting the individual sites—always respect private property and observe all signs and warnings.

Introduction

The outermost layer of planet Earth, its crust, is composed of rock, a substance described as an aggregate of one or more minerals. A mineral, in turn, is commonly defined as a naturally occurring, inorganic solid that has a definite arrangement of atoms and a characteristic chemical composition.

Geologists have identified more than four thousand minerals but have not reached consensus on how many different types of rocks there are. In reality, with permutations of the number of minerals, an almost infinite number could exist. Mineral samples collected an inch or so apart in the very same rock exposure might differ in crystal or grain size, chemistry, mineral arrangement, and other distinguishing aspects such as density, porosity, and color.

Rocks are classified into three groups, based on how they formed: igneous, sedimentary, and metamorphic.

Igneous rocks, the first to form on Earth, solidify from a molten state. Molten material beneath the surface—magma—crystallizes slowly to form coarse-grained, intrusive igneous rock, such as granite and gabbro. Magma that erupts onto the surface—lava—cools rather quickly to form fine-grained, extrusive igneous rock, such as basalt and rhyolite.

Exposed to the atmosphere, rock slowly disintegrates into particles of various sizes, termed sediment. Moving water, ice, wind, ocean currents, or gravity move sediment from its point of origin to a site of deposition. There, either pressure or cementation with a natural mineral cement compacts the sediment into sedimentary rock.

The most common types of sedimentary rock are conglomerate, which is composed of particles larger than sand grains; sandstone, compacted sand grains; shale, a form of indurated mud; and limestone, an assemblage of microscopic shells. The precipitation of calcium carbonate from freshwater and salt water also forms limestone.

Whenever igneous or sedimentary rock is buried deep within Earth, becomes involved in the dynamics of mountain building, or comes into contact with either magma or lava, it undergoes physical and chemical change. Great pressures and intense heat rearrange preexisting mineral and texture characteristics into those that define a metamorphic rock, such as schist, gneiss, and marble. Exceedingly high temperatures and pressures cause metamorphic rock to melt into magma. The sequence of events involving rock formation and change, from magma to igneous to sedimentary to metamorphic rock and back again to magma, is called the rock cycle.

RELATIVE AGE DATING

Established only about 235 years ago, the science of geology is a relatively youthful arena of study. Long before this, however, dating back to even the dawn of analysis, humans wondered about their planetary home. Questions abounded: How are mountains built? Why are shells encased in rock and found miles from the nearest sea? What causes the land to periodically shake? Why do volcanoes erupt? And how old are rocks? Indeed, how old is Earth itself?

By the midpoint of the sixteenth century, tentative steps had been taken toward the understanding of Earth processes. Georgius Agricola published his thoughts on the power of water and wind as predominant forces of erosion, the role groundwater plays in cementing sand particles together, and the link between subterranean heat and volcanism. He was an early master of applied geology, which looks at how geology affects humans.

During the early decades of the seventeenth century, two forward-thinking individuals paved separate paths toward new schemes of unconventional thinking. Robert Hooke, an Englishman, showed an enlightened awareness of earthquakes and their role in land elevation and subsidence. Nicolaus Steno, a Dane, argued that mountains could be formed by one of three means: uplift, volcanism, or erosion. These were men who felt comfortable with multiple working hypotheses. Of greater significance, Steno applied his observations of river flooding and sediment transport to the shaping of several principles that form the bulwark of modern geologic study:

- The principle of superposition: In a vertical sequence of undisturbed, layered rock, the oldest layer is at the base and the youngest on top.
- The principle of original horizontality: Sediment is initially deposited in a flat, horizontal layer.
- The principle of lateral continuity: A layer of sediment will extend in all directions until terminated by thinning or gradation into another type of sediment.

Step-by-step scientific inquiry was advancing. In time, three other principles were added to the list:

- The principle of crosscutting relationships: An igneous intrusion or fault is younger than the rocks it intrudes or cuts across.
- The principle of faunal succession: Fossil assemblages succeed one another in a regular and determinable order.
- The principle of inclusions: Fragments of rock in a layer of sediment are older than the layer in which they are contained.

These time-honored principles are still in use today whenever geoscientists attempt to read a previously undiscovered exposure of rock.

As the eighteenth century matured to old age, Neptunism, the belief that the great majority of rocks precipitated from a universal ocean, became all the rage. Those of an opposing conviction, the Vulcanists, subscribed to the idea that lava flows and related crystalline rocks resulted from the cooling of once-molten material, with or without the presence of seawater. And then there were the Plutonists, who advanced the concept that rock is formed when subterranean molten masses solidify.

While some investigators studied how rocks form, others pondered how the surface and complexion of the crust had come to be, and how it might have changed over the years. In 1650, James Ussher, Archbishop of Armagh, Ireland, announced that Earth had been born on Sunday, October 23, 4004 BC, a date based on genealogical information contained in his Bible. This declaration became the foundation for two beliefs that controlled scientific investigation for decades: Earth was but 6,000 years old, and its surface was constantly being altered by processes of catastrophism—sudden, violent, short-lived, worldwide disasters produced by unknown causes that no longer operate.

Catastrophism reigned supreme until 1788, when James Hutton, the father of modern geology, advanced the doctrine of uniformitarianism, the concept that the forces of nature in operation today differ neither in kind nor energy from those that operated in the past. This concept is abbreviated today as "The present is a key to the past." Geology was coming of age.

By the dawn of the nineteenth century, geologists had gathered a tremendous volume of field data, but it was of minimal value unless it could be organized in chronological order. They needed a time chart—a listing of the chapters of Earth history in sequential order. The ideal model would embrace all of geologic time and be applicable from one continent to another.

After decades of debate, the scientific community reached a consensus. The time-honored principles of superposition, original horizontality, lateral continuity, crosscutting relationships, faunal succession, and inclusion could be employed to separate sequences of rock into systems, each representative of a particular episode of Earth history. An arrangement of these systems into a proper and orderly sequence quickly became the first geologic column, in effect a relative timescale.

Today, the 4,600-million-year history of Earth is subdivided into an arrangement of eons, eras, periods, and epochs. An eon, the greatest expanse of time, is made up of two or more eras, units originally defined by the absence or presence of life. In turn, each era is composed of multiple periods that are characterized by episodes of significant mountain building or changes in life-forms or which life-forms dominated. An epoch is the smallest commonly recognized unit of time.

The geologic column affords geologists a form of verbal shorthand useful in the discussion of any aspect of Earth history. They can, for example, speak of the Devonian period, knowing it occurs after the Silurian and before the Carboniferous in the sequence of time, and any and all rocks that came into being during that time are assigned to the Devonian period.

Early on, it became clear that this means of communication had but one flaw—it designated only the chronological order of relative time. The advance of technology was highlighting the importance of quantifying the timescale so that absolute dates bracketed each unit of time. With such an adaptation, inquiry and reasoning would enter a new era of understanding. That era began with the discovery of radioactivity in 1896.

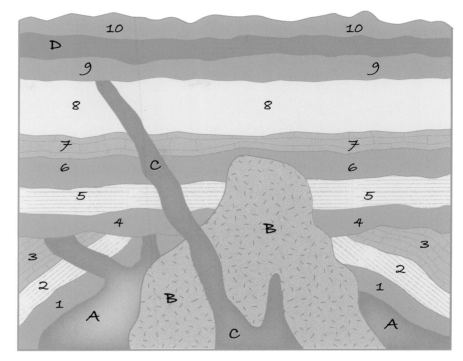

The sequence of geologic events associated with this hypothetical exposure of rock, using several of the six principles that form the tool kit of a field geologist, is listed below.

- Sedimentary rock 1 is the oldest, followed in relative age by sedimentary rocks 2 and 3 (principle of superposition).
- Igneous intrusion A is younger than rock 3 (principle of crosscutting relationships).
- Rocks 1 through 3 have been folded and no longer lie in their original horizontal position, indicating that a period of mountain building occurred after 3 was deposited, disturbing the sequence of rock (principle of original horizontality).
- Segments of intrusion A and the uplifted portions of rocks 1 through 3 terminate at the base of rock 4, indicating a period of erosion followed the episode of mountain building (principle of lateral continuity).
- Sedimentary rock 4 was deposited next, followed by rocks 5 through 8 (principle of superposition). Sometime during this episode of deposition, igneous intrusion B cut across intrusion A and invaded rocks 4, 5, and 6, evidence that these three layered units are pre-B in age (principle of crosscutting relationships). Because intrusion B does not extend into rocks 7 and 8, their ages relative to B cannot be determined. Rocks 7 and 8 are younger than 6 (principle of superposition), but their relationship to intrusion B can be stated only as pre-B or post-B.
- Intrusion C cuts across intrusion B, as well as sedimentary rocks 4 through 8. Intrusion C, the youngest of the three intruded events, is younger than rock 8 (principle of crosscutting relationships).
- Intrusion C terminates at the base of sedimentary rock 9, evidence of a period of erosion that occurred post-8 and pre-9 in time.
- Sedimentary rock 9 was deposited after the period of erosion and sometime thereafter buried by lava flow D (principle of superposition).
- Sedimentary rock 10 was deposited after lava flow D. Its age can be simply stated as post-D and pre-today.

EON/ERA		PERIOD		EPOCH	AGE today
PHANEROZOIC	CENOZOIC	QUATERNARY		HOLOCENE	.0117
				PLEISTOCENE	2.6
		NEOGENE		PLIOCENE	5
				MIOCENE	23
		PALEOGENE		OLIGOCENE	34
				EOCENE	56
				PALEOCENE	65
	MESOZOIC	CRETACEOUS			145
		JURASSIC			199
		TRIASSIC			251
	PALEOZOIC	PERMIAN			299
		CARBON-IFEROUS	PENNSYLVANIAN		318
			MISSISSIPPIAN		359
		DEVONIAN			416
		SILURIAN			443
		ORDOVICIAN			488
		CAMBRIAN			542
PRECAMBRIAN	PROTEROZOIC				2,500
	ARCHEAN				4,000
	HADEAN				4,600

The units of geologic history shown in this timescale are used throughout this guide. Radioactive age dating procedures determined the ages listed in the right-hand column, which are given in units of millions of years.

ABSOLUTE AGE DATING

Once the geologic timescale had been devised on the basis of relative age—"this rock is older than that rock"—the obvious next step was to find a way to assign absolute ages to not only rocks, but also the many subdivisions that compose the geologic timescale. Many scientists still believed Earth was some 6,000 years old, but others thought otherwise. Through the years, various attempts to quantify time had concentrated on natural processes that were supposedly rooted in scientific evidence rather than any degree of theological disclosure. These attempts—documentation of the ingenuity of mankind—and the related "age of the Earth" results include:

- Heating iron spheres and recording the time they took to cool to room temperature, and extrapolating to a sphere the size of Earth: no less than 75,000 years.
- Calculating how long it had taken the sun to evolve to its present diameter and brightness: 20 to 25 million years.
- Determining the relationship between modern rates of sediment deposition and the aggregate thickness of the worldwide sedimentary rock record: 55 million years.
- Dividing the total load of dissolved oceanic salts by the annual input from wind, ice, and river systems: 100 million years.

Each of these age dates were proven wrong, but the fact that the analyses suggested Earth was much older than 6,000 years remained a significant contribution to the understanding of geologic time. Then, when scientists unlocked the secrets of the atom and learned more and more about how protons and neutrons are bound together in the atomic nucleus, the door opened to a new avenue of investigation.

Radioactive elements have been a part of the physical world for as long as Earth has been rotating around the sun. From day one of their formation, these elements are chemically unstable, a state that has no permanence in nature. Since stability is the sought-after condition, all radioactive elements are gradually transformed from the unacceptable to the desired condition, a process called radioactive decay.

Different radioactive elements have different decay rates, but for any one particular element the rate is constant and independent of changes in pressure and temperature, altitude or depth of burial, chemical or physical state, or association with any combination of other elements. This independence provides a type of reliable "atomic clock" by which the age of rocks—and by extension the age of the Earth—can be quantified.

In 1907, Bertram Boltwood, an American chemist, suggested the metal lead was the stable end product that resulted from the multiple-step process involved in the decay of uranium, a common radioactive element. After he had dated a dozen or more rock samples using this uranium-lead process, he announced Earth's age as lying somewhere between 400 million and 2,200 million years.

Arthur Holmes, an English geologist, continued the work of Boltwood and spent his entire life perfecting radioactive age dating technology. At the time of his death in 1965, several months after the release of the second edition of his seminal *Principles of Physical Geology*, he was still pondering the antiquity of Earth. In that book he records his final thoughts on the subject: "We may therefore conclude that in round figures the age of the earth's crust lies between 3,500 and 5,500 million years."

Today, almost five decades after Holmes' death, his halfway figure of 4,500 million years is surprisingly close to the true mark of 4,600 million years. Several rock samples gathered from beyond the confines of Earth verify this age. The oldest of the rocks collected on the moon is a 4,500-million-year-old specimen the Apollo 17 crew returned. In addition, standard radiometric techniques have revealed that scores of meteorites lie between 4,500 and 4,600 million years of age. Earth is, indeed, a senior citizen.

The quest to identify the oldest rock has not been an easy task. More than 90 percent of the world's continental rock is younger than 2,500 million years, and oceanic rock is even younger. It is questionable whether the "Adam" of solid Earth material still exists. Involvement in the rock cycle may have melted it, or perhaps metamorphism has altered it to a new state.

For years, bedrock forming the basement of southwestern Minnesota held the American record, but today the winning date—as old as 3,960 million years—belongs to an exposure of Quad Creek quartzite along the Beartooth Highway, which leads to the northeastern entrance to Yellowstone National Park. In 2008, a team of geologists announced they had uncovered an expanse of rocks from the eastern shore of Hudson Bay that dates from 3,800 to 4,280 million years in age, the latter the oldest date ever reported for any rock specimen from any region of Earth. Is there any terrestrial material older than this Canadian rock? The answer is unexpected. Recently geologists found zircon minerals enclosed in a 3,000-million-year-old gneiss collected from Western Australia to be 4,404 million years old, a date less than 200 million years distant from the very birthing moment of Earth. Maybe tomorrow that "day one" rock will be discovered.

PLATE TECTONICS THEORY

The ideas behind plate tectonics have been around for a long time. As soon as maps of the various landmasses became available with shapes, sizes, and coastlines akin to reality, individuals who spent time studying them became intrigued by the configuration of Africa, South America, and North America. The coastline of each one seemed to be related to the coastlines of the other two. Most specifically, the eastern protuberance of South America appeared to have been created for the sole purpose of eventually being fit into the curve of western Africa. It was as if these continents had once been one and then were torn apart by some Earth-rending force.

Abraham Ortelius, acknowledged as the editor of the first modern atlas, suggested as early as 1596 that consideration be given to the "projecting parts" of Europe and Africa. In the decades that followed, other prominent men added their thoughts to the developing theories that, far from being an immobile planet, Earth is continually being altered by regenerative processes. For example, Sir Francis Bacon suggested in 1620 that the mapped coastlines of South America and Africa were "no mere accidental occurrence." Benjamin Franklin thought the fossils he collected in the high hills of Derbyshire, England, in 1782, had reached elevated heights because Earth's crust was "floating on a fluid interior" and thus was "capable of being broken and disturbed." During his five-year exploration of South America, beginning in 1799, Alexander von Humboldt recognized that the mountainous terrain ending at the continent's eastern shoreline appeared to continue beyond the vast Atlantic Ocean and onto the western edge of Africa. He was one of the first to suggest that "matching edges" be combined with "geologic fabric" to support the concept of a once-singular landmass

These early ideas had one common thread: the obvious fit of the continents had something to do with the Noachian flood, the catastrophic event in biblical history that supposedly altered the face of the planet. In 1858, however, French naturalist Antonio Snider-Pellegrini presented a refreshingly different idea. He argued that the biblical flood was the

The geometric fit of the continents of North America, South America, and Africa was an early indication that these landmasses were once joined as one.

result, not the cause, of continental breakup. Soon, the rush was on to find fossil and rock evidence to support the processes of continental rupture and displacement—often with astounding results. For example, in 1885 Austrian geologist Edward Seuss presented evidence that similar fossil plant forms could be found in South America, India, Australia, Africa, and even Antarctica. Convinced these landmasses were once one, he named that hypothetical mother continent Gondwanaland.

With support for the idea of a universal flood waning, and continental fit no longer considered a coincidence, in the latter half of the nineteenth century the scientific community was finally set free from the restraints of theological approval. More than 250 years had passed since Ortelius published his maps. It was time for a new and succinct approach to the understanding of global geologic process.

During the 1908 meeting of the Geological Society of America, Frank Taylor, a geomorphologist with special interest in glacial landforms, argued against conventional concepts regarding global geologic process. He methodically presented his thesis: "A great world-belt . . . of Tertiary fold-mountains almost circling the Earth" had been formed by massive lateral forces over an "extended period of time," the result of a "mighty creeping movement" directed by

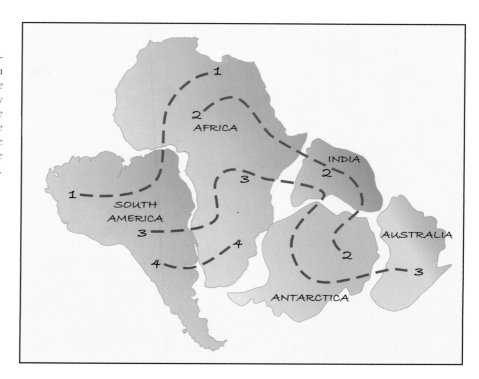

The presence of similar fossil species (1 through 4) in various southern hemisphere continents, separated by vast oceans, makes sense when the continents are joined together into a single landmass according to the precepts of continental drift.

crustal fragments moving "from polar to equatorial latitude." Referring to a crude bathymetric chart, he became the first geologist to identify the Mid-Atlantic Ridge as a "submerged mountain range of a different type," one that marked "the original place of the great fracture" that "remained unmoved whilst the two continents on opposite sides have crept away in nearly parallel and opposite directions."

These thoughts are believed to be the very first discussion of the processes that six decades later would become the scientific framework of the seminal, universally accepted theory of plate tectonics. What the theory of evolution did for the understanding of life and the discovery of radioactivity did for the comprehension of the atomic world, the theory of plate tectonics did for a better understanding of Earth processes, by assembling previously discombobulated geologic data and binding them together in a unifying framework of understanding.

This revolutionary theory can be presented with four basic concepts: (1) The crust and the uppermost region of the mantle of the Earth are composed of a series of rigid plates that fit together like the plates that make up a turtle's shell. (2) These plates are in slow but constant movement relative to each other, at a rate about as fast as the human fingernail grows. Some collide, others move apart, and a few grind against each other. (3) Most of the world's geologic activity, such as mountain building, volcanic eruptions, and earthquakes, takes place along the edges of the plates. (4) Geologic activity is minimal in the interior regions of the plates.

The theory of plate tectonics explains the distribution of the plots of seismic and volcanic activity; the global array of ancient reef, coal, and fossil deposits; the geometry and evolution of the continents; and the location of folded and faulted mountain ranges. It is extremely valuable in identifying areas most apt to contain fossil fuel reserves versus regions more adaptable to the formation of mineral deposits. No longer is the Earth viewed as merely a combination of continents and ocean basins. It is, instead, a mosaic of major and minor plates that are constantly interacting in a time continuum affecting the very essence of planetary process, ranging from the evolution of flora and fauna to the rise and fall of mountains and ocean levels.

The birth of plate tectonics can be traced to the 1915 publication of *The Origin of Continents and Oceans*, by a then-obscure German meteorologist named Alfred Wegener. Universally acknowledged today as the originator of the theory of continental drift, the precursor to the plate tectonic

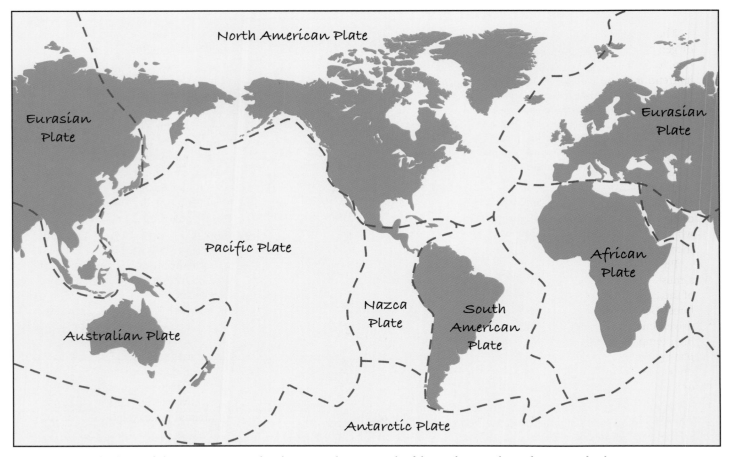

The theory of plate tectonics posits that the crust and upper mantle of the Earth are made up of a mosaic of eight major plates (shown) and six or more minor plates (not shown). Zones along which the plates are moving apart, coming together, or grinding past each other identify their boundaries.

theory, Wegener was a Renaissance man whose thoughts ranged far beyond wind, rain, and sunshine. He was the first to compile the myriad of Earth-process thoughts that had arisen over the centuries into a compact text. He was also the first to argue that all the continents had been joined together at one time as a supercontinent. He named the supercontinent Pangaea, meaning "all Earth," and said it had been surrounded by a universal ocean he called Panthalassa.

His book was widely read and generated much opposition. One major point of contention dealt with the "force" behind his ideas of drifting continents. Since he wasn't sure what that might be, he attempted to answer by analogy: imagine lightweight continents sailing through a denser crust like icebreakers plowing through a frozen ocean, while periodically they break up and drift apart. The reaction was immediate and ranged from "a footloose idea" to "utter, damned rot." When he died on the Greenland ice cap in 1930, his beliefs were temporarily buried with him.

After the conclusion of World War II, geologists undertook the gargantuan task of converting tens of thousands of miles of collected ocean-sounding profiles into a worldwide catalogue of bathymetric charts. From this effort emerged the first realistic, three-dimensional representation of the topography of the ocean floor. Early in this process, cartographers recognized a sharply defined ridge in the Atlantic Ocean, halfway between Europe and North America and halfway between Africa and South America.

The existence of segments of this ridge had been part of general knowledge for decades, but this new information fully identified the Mid-Atlantic Ridge and extended its trend into the Pacific, Indian, and Arctic regions in the form of a 40,000-mile-long, world-encircling mountain chain that contains the highest peaks—as measured from the ocean floor—known on the face of Earth. Along with the oceans and continents, the ridge, once discovered, was recognized as one of the most prominent features defining the surface of the planet. And so the modern era of oceanographic research had begun. Each new discovery quickly prompted a series of questions. By the mid-1960s, volumes of new data updated the once-rejected ideas of Wegener, which ultimately were repackaged as the plate tectonics theory.

Post–World War II research that revealed the presence of a mushy, semiliquid zone confined to the uppermost layer of the Earth's mantle resolved Wegener's question of driving force. Geologists today believe that convection cells, heat-driven currents of semimolten rock fueled by the extreme high temperatures of the Earth's core, are the force behind plate movement. As these rising currents approach the underside of the crust, they spread apart and move laterally, dragging the overlying plate along much like a conveyor belt transports its load of material. The currents then slowly cool, become dense, and finally sink back into the base of the mantle, where they are reheated and begin the journey upward again.

The rise and fall cycle of convection cells produces three categories of plate movement: divergent, where plates move apart; convergent, where plates slowly collide; and transform, where plates grind against each other in opposing directions.

Let's start with divergent boundaries. Real-time data from deepwater cameras used to study the development of the Mid-Atlantic Ridge have been correlated with other sources of information to show that the Atlantic Ocean was born through a mechanism called seafloor spreading. As rising convection cells diverge from one another, they cause the tectonic plate above to fracture. Magma oozes into the fracture, slowly crystallizes, and temporarily heals the wound. Soon, however, fracturing begins again, and more divergence takes place, along with more magma injection. In this manner, the entirety of the Atlantic Ocean continues to grow in width, on average 1 to 2 inches per year, as seen by the eruptions of lava that periodically take place on the islands of Iceland and Tristan da Cunha, which are part of the ridge.

Not all divergent boundaries are as old as the Mid-Atlantic Ridge. Along the East African Rift, divergence is in a very early stage of development. There it appears that a portion of the African Plate is splitting into two smaller plates. Further north, the Red Sea represents a more advanced state of divergence. Some divergent boundaries never grow beyond the youthful stage—meaning they never become ocean basins. The Midcontinent Rift, a structure of Precambrian age that underlies the waters of Lake Superior, is a case in point. Born some 1,100 million years ago, it ceased to be active after a mere 22 million years, when the collision of proto–North America with a smaller landmass named Grenvillia altered its tectonic character from divergence to convergence, closing the rift.

New rock is constantly being generated along divergent boundaries. However, because the surface area of Earth has remained essentially constant throughout time, new-rock volume must somehow be balanced by the consumption of an equal volume of rock elsewhere. Convergent boundaries are, by definition, the zones along which rock is consumed by the mantle.

Although all convergent zones are basically the same in their overall mode of operation, the specifics of plate collisions are determined by the nature of the involved plates. Convergence can take place between two oceanic plates (composed of seafloor rock), two continental plates (composed of continental rock), or one of each type.

The convergence of two oceanic plates has created some of the great island trends of the world. When oceanic plates collide, the plate with the higher density will plunge, or be subducted, beneath the other. At a depth of about 60 miles in these subduction zones, the leading edge of the descending plate will partially melt and form a pocket of magma that will rise slowly to the surface and erupt, forming an island arc, a linear chain of volcanic islands. Interesting examples exist throughout the northern and western portions of the Pacific Ocean, including the Aleutian, Japanese, and Philippine island arcs. The eighteenth-century eruption of Mount Fuji and that of Mount Pinatubo in 1991 are evidence that these islands are indeed volcanic.

When an oceanic plate collides with a continental plate, the denser oceanic slab is always subducted beneath the lighter continental mass. Magma is still produced as the oceanic plate melts, producing chains of volcanoes on the over-riding continental plate. Prime examples include the Andes mountain chain, created by the continuing collision of the Nazca Plate with the South American Plate, and the Cascade

Range of Washington and Oregon, the result of convergence of the North American Plate with the Pacific Plate. The eruption of Mount St. Helens in 1980 is evidence that the Cascade Range is still growing through plate tectonic activity.

Continent-with-continent convergence has created some of the largest mountain ranges in the world. Because the colliding edges of both plates are composed of relatively low-density material, making them quite buoyant, neither one is subducted. The result can be a fender-bender collision of humongous proportion. The Himalaya Range, home to more than one hundred peaks that exceed 23,000 feet in elevation and Mount Everest, the highest mountain in the world, is a textbook example. Others are the Alps of Europe and the Atlas Range of North Africa. Closer to home, continent-to-continent collisions between early versions of the African, North American, and Eurasian plates formed the Appalachian Mountains over a period of several hundred millions of years, during the Paleozoic era.

The third type of plate boundary takes place where two plates, either of a continental or oceanic nature, grind past each other in a lateral movement known as transform motion. While volcanism is generally absent within these zones, they are home to earthquake belts of international fame. The best example is the San Andreas Fault, a very active zone of seismic activity caused as the Pacific Plate grinds past the North American Plate.

Before the years of World War I, the Atlantic Ocean was known to the extent of only one depth sounding for every 5,400 square miles, and the Pacific to only one for every 10,000 square miles. Today, practically every segment of the world's oceans has been depth sounded and age dated. The oldest ocean floor found anywhere, in the western Pacific Ocean, is some 180 million years old. This age is but 4 percent that of the oldest known continental rock, a stark indication that Earth's continents are not recycled and its seafloors are. Throughout geologic time, continents have grown by accretion—increasing their size through collision with island arcs and other microcontinents. In contrast, oceans continue to open and close, the victim of plate boundary divergence and convergence.

As soon as the principles of the theory of plate tectonics were accepted, geologists began collecting additional information to test the theory, hoping to build an even stronger foundation to this concept of Earth process. Earthquakes provided some of this new information.

For years seismologists had been interested primarily in plotting the worldwide distribution of earthquake epicenters, those geographic points vertically above the heart,

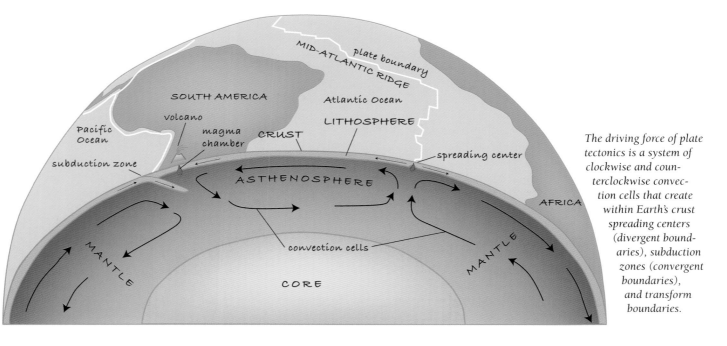

The driving force of plate tectonics is a system of clockwise and counterclockwise convection cells that create within Earth's crust spreading centers (divergent boundaries), subduction zones (convergent boundaries), and transform boundaries.

or focus, of Earth tremors. When they began to study the relationship between focus and epicenter, they quickly recognized an interesting pattern. For example, along the convergent boundary forming the western border of the Pacific Plate, in the western Pacific Ocean, shallow-focus quakes occurred immediately below the deep-sea trench that marks the surficial contact of the Pacific and Eurasian plates. In contrast, intermediate- and deep-focus earthquakes occurred progressively farther to the west, toward the shoreline of China. In essence, the plots painted a picture for seismologists. Clearly, the Pacific Plate was the denser of the two and plunged westward underneath the Eurasian Plate in a subduction zone, generating earthquakes with ever-deeper origins.

Deep-sea drilling from the decks of specially designed oceanographic vessels provided information that also supported the theory of plate tectonics. Comparing the oldest ocean floor sediment from each drill site to its distance from the nearest active oceanic ridge (spreading center) revealed that the sediment age increased with increasing distance from the ridge. Thus, the most juvenile ocean crust was adjacent to the axis of the oceanic ridges, where new crust was being or had most recently been created, and the oldest was along the subducted margins of the plates.

Radiometrically age dating rocks from the Hawaiian Island–Emperor Seamount chain of volcanic islands in the Pacific Ocean showed that age increased with increasing distance from the still-growing island of Hawaii, the youngest of this several-thousand-mile-long chain. Geologists proposed that this linear relationship was due to a hot spot, a rising plume of magma, beneath the Pacific Plate. As the Pacific Plate moved northwesterly over this site of volcanic activity, periodic eruptions created new volcanic islands. Hawaii is currently positioned over the hot spot, which is why it is so volcanically active.

Finally, the most startling new information in support of the theory of plate tectonics is associated with the 1977 discovery of black smokers along the crest of the spreading-center ridge in the eastern portion of the Pacific Ocean. Subsequent discoveries show that these erupting hot springs—a mixture of boiling water, hydrogen sulfide, and other gases under great pressure that are in reality the exhaust of the thermal energy that drives ocean spreading—occur exclusively along divergent boundaries. The more than three hundred new species of life found living in close association with these vents are dependent not on the energy of the sun, but on the chemicals dissolved in the smoker fluids. Some scientists have suggested that black smokers once served as the very incubation sites for life on Earth.

A SHORT HISTORY OF THE EARTH

The theory of plate tectonics has been an integral part of the evolution of scientific thought for almost fifty years. Wherever possible, I incorporate plate tectonic interpretations relating to the geologic story of sites contained in this guide. The following abbreviated Earth history will help you better understand the geologic events associated with each individual geo-site.

HADEAN EON:
4,600 to 4,000 Million Years Ago

Perhaps the Hadean eon should not be considered a part of the geologic record since rock-solid data about this time period does not exist. Even the name is controversial, derived as it is from Hades, the Greek god associated with death, to describe an eon that centers on birth. Regardless of the nature of the debate, there is a consensus within the scientific community that by the end of this 600-million-year episode a proto-Earth had been formed—complete with a geosphere, atmosphere, and hydrosphere.

The eon began when a cloud of gas and dust born from the explosion of a supernova, an old star of massive size, shrank to form proto-Earth, which early on assumed a molten state, the product of both frictional heat and heat created by radioactive decay. Heavy material sank to the center to form an ironlike core, around which lighter elements rose to the surface as a low-density froth. Material caught between the two became Earth's mantle. The most volatile molecules—carbon dioxide, nitrogen, ammonia, hydrogen, methane, and water vapor—bubbled above the surface as a primitive, toxic atmosphere. Eventually, isolated blocks of rock crystallized on the surface to become the first protocontinents. But this situation did not last for long. Meteors blasted this primitive crust into fragments that were quickly recycled into the molten depths by circulating heat cells.

In spite of these setbacks, geologic progress was under way, and in time a permanent crust began to clothe Earth. Temperatures declined to the degree that water-laden clouds produced the first storms. Oceans began to fill in low-lying topography, and river systems were established. Finally, the water cycle was initiated, along with the processes of weathering and erosion.

ARCHEAN EON:
4,000 to 2,500 Million Years Ago

In the beginning, Earth's internal heat flow was three times that of today. As meteor bombardment finally subsided, rampant volcanoes spewed forth volumes of lava that reached 3,000 degrees Fahrenheit, compared to lava today, which forms from rock that liquefies at 2,000 degrees Fahrenheit. The Earth was seething, rambunctious, and undisciplined during the Archean eon. Still, surface temperatures were cool enough for rocks to crystallize and continental crust to grow. Archean rocks make up as much as 7 percent of the combined landmasses of North and South America, Africa, Eurasia, and Australia and form the very geologic hearts, or shield areas, of these continents.

Archean terranes in Greenland, Labrador, and the states of Wyoming and Minnesota indicate this was the time when the structural framework of North America came into being. Formed at disparate localities, these regions were later sutured together by plate tectonic processes. Volcanoes continued to vomit vast volumes of toxic fumes and liquids. Extensive fields of bun-shaped pillow lava, formed when lava extrudes into bodies of water, testify to the presence of oceans. Since little if any carbonate rock dates to this time, these waters were probably quite acidic. Banded iron formations—a worldwide source of iron ore—appeared 3,600 million years ago, with a mineral content suggesting they formed under oxygen levels much lower than those prevalent today, further evidence of a noxious Archean atmosphere.

Life—the animated, indispensable state of vitality and being—appeared upon the geologic stage. Indeed, the Archean eon is often called the Age of Prokaryotes, referring to single-celled forms of blue-green algae that lack a true nucleus and reproduce asexually. The algae trapped sediment and built moundlike structures known as stromatolites, and these fossilized remnants occur in rocks that are as old as 3,500 million years. Because they were built by photosynthesizing organisms, the stromatolites represent perhaps the most significant transition in Earth history: from an oxygen-starved to an oxygen-rich atmosphere. Earth would never again be the same.

PROTEROZOIC EON:
2,500 to 542 Million Years Ago

The account of the Proterozoic eon is Earth's greatest chapter, containing a full 42 percent of its history and a time of mind-boggling accomplishment and unprecedented variance. Consider the topics: atmospheric conversion, supercontinent birth and death, universal glaciation, mass extinction, and the invention of sex. The beginning of this era 2,500 million years ago was near the central point of Earth history. Was a planetary midlife crisis under way?

By mid-Proterozoic time the dynamics of plate tectonics, whereby continents periodically come together and then are torn apart while oceans are born and then die, were near perfection, with one exception. Approximately 1,100 million years ago, a thermal plume from a source deep in the Earth lifted and fractured the subcontinent of Laurentia (proto–North America), and in the process created the Midcontinent Rift that extends from Kansas to Ohio via Lake Superior. This master rift grew in width for 22 million years until it was closed by one of the collisions that formed the ancestral supercontinent Rodinia.

Beginning some 700 million years ago, Rodinia started to break into subparts, creating shifts in weather and erosion patterns that, along with reductions in greenhouse gases, caused global temperatures to decline enough for glacial ice fields thousands of feet thick to extend from pole to pole for millions of years. During this time as much as 95 percent of aquatic life may have become extinct on what some geologists call Snowball Earth.

More than 100 million years before this happened, however, rapidly rising levels of atmospheric oxygen sent a signal to the simple, asexual life-forms then in existence: evolve or die. Most found this "oxygen catastrophe" too much to survive, but a few adapted by forming symbiotic associations with each other. The organisms that resulted were on average ten times larger than their predecessors, sported a distinct nucleus, and exchanged genes sexually, producing with each generation different genetic configurations and biologic variation.

By the end of Precambrian time, an informal term comprising the Hadean, Archean, and Proterozoic eons, Earth had undergone 88 percent of its history, the continental cores were complete, North America had grown to approximately 75 percent of its present size, oxygen levels were still rising, and the sexual revolution was well under way. The Cambrian explosion lay dead ahead.

CAMBRIAN PERIOD:
542 to 488 Million Years Ago

The 4,600-million-year history of Earth is divided into four eons. Besides the three that form Precambrian time, there is the Phanerozoic eon. The Phanerozoic extends from 542 million years ago through today, covers 12 percent of geologic history, and is characterized by a worldwide distribution of sedimentary rock that is minimally altered, widely accessible, often fossil bearing, and ultimately easier to interpret than heavily altered Precambrian rock. Geologists, therefore, are able to focus on smaller subdivisions of time. The Phanerozoic comprises three eras—Paleozoic, Mesozoic, and Cenozoic—that are further subdivided into periods, which is the unit of time I use to focus the rest of this discussion of geologic time.

The opening chapter of the Paleozoic era emphasizes the Cambrian explosion, an abrupt appearance of a wide variety of large, multicelled, skeletal life-forms, including representatives of every modern animal phylum with a three-part anatomy—head, body, and tail—that coordinated the functions of sense, energy, and motion.

Four ancestral landmasses, all derived from the disintegration of the supercontinent Rodinia, comprised the principal geography of Earth: Baltica, today eastern and northern Europe; Siberia; Laurentia, or proto–North America; and Gondwanaland, now Africa, India, Australia, New Zealand, South America, and Antarctica. A sudden acceleration in plate motion shifted the geographic positions of at least two and perhaps all of these landmasses as much as 90 degrees in latitude. Laurentia moved from the south pole region to near the equator at a rate measured in feet per year, compared to the norm of inches per year. The Iapetus Ocean occupied the area between the landmasses of Laurentia and Baltica. Climates warmed and oxygen levels continued to climb. Marine fauna experimented with the synthesis of collagen on their way to forming bone. Innovative processes of evolution were under way.

ORDOVICIAN PERIOD:
488 to 443 Million Years Ago

While the 1840 Presidential election of William Henry Harrison—hero of the 1811 Battle of Tippecanoe—and his running mate John Tyler is remembered for the campaign slogan of "Tippecanoe and Tyler too," nobody *remembers* the Ordovician period. Humans were still nearly half a billion years from their earliest memories. However, the period could be succinctly described with a twist on that campaign slogan: "Tippecanoe and Taconica too."

Positioned between 25 degrees north and 25 degrees south latitude, Laurentia had a very low topographic profile that was exposed to extensive erosion. The widespread, tropical Tippecanoe Sea, an extensive shallow portion of the Iapetus Ocean that flooded Laurentia during this time, covered an area from Pennsylvania west to Nevada and north to Hudson Bay, an event often described as the greatest-ever submergence of any continental landmass. Because Laurentia straddled tropical latitudes and was without highlands, limestone was the dominant rock deposited, often with more than 75 percent shell material.

Surviving Cambrian species competed with new forms of coral, bivalve life, snails, eel-like conodonts, and myriad jawless fish, the first true vertebrates. Filter feeders and surface feeders replaced bottom-feeding types, while the number of extant families increased threefold, adding to the complexity of the food chain. Bleak land became more pastoral in appearance. Proterozoic eon algae evolved onto the land as primitive vegetation in steps, from saltwater to brackish to freshwater environments.

Beginning some 470 million years ago, submarine avalanches carried large boulders into calm water areas along the eastern seaboard of Laurentia. Volcanic ash from eruptions settled onto the ocean floor and was interbedded with carbonate sediment, while deeply buried limestone was transformed to marble due to high temperature and pressure. Subterranean pods of magma, formed in subduction zones, migrated to the surface, creating Taconica, an arc of volcanic islands that rose to heights of 15,000 feet and then collided with Laurentia. This collision resulted in the Taconic orogeny, the first of three mountain building episodes that eventually gave world-class status to the Appalachian mountain chain.

SILURIAN PERIOD:
443 to 416 Million Years Ago

Laurentia settled into an episode of tranquility during the Silurian period. The atmosphere was balmy and somewhat arid, glaciation was confined to the south pole, and volcanic activity declined considerably. Deposition of sediment, eroding off the declining heights of Taconica, added to the ever-increasing mass of North America. In the Tippecanoe Sea, colonies of coral built extensive reef complexes, creating basins in which seawater evaporated, forming thick deposits of salt.

A diverse group of 10-foot-long sea scorpions preyed upon the masses of brachiopods that made up 90 percent of Silurian marine life, while other new species menaced anything that moved. Through the process of evolutionary opportunism, which is the specific pattern of development in the history of any species, the acanthodian class of fish evolved. These fish were characterized by a movable jaw, an extended spine, and body-covering scales.

Even as the marine food chain was becoming longer and the planet's total biomass larger than ever, the big news centered on the development of land-based flora. Vascular plants with a developed system of roots and leaves brought an organic softness to the landscape and opened the door for amphibians. The onset of the Acadian orogeny, a mountain building event that sculpted the landscape from Newfoundland to Virginia, ushered out the Silurian period. The global dance of tectonic plates was now in full frolic.

DEVONIAN PERIOD:
416 to 359 Million Years Ago

During the Devonian period Laurentia merged with other northern hemisphere microcontinents, giving rise to a new amalgamation of land called Laurasia. The Iapetus Ocean, which separated Laurentia and Baltica, was further reduced in extent during this process, and a new balance of global land geography developed in the form of two main masses: Gondwanaland south of the equator and Laurasia to the north. The Acadian orogeny was ending, having created the Acadian Mountains, the heights of which are reflected in the mile-thick sequence of sedimentary rock that underlies New York State. These rocks formed from the weathering and erosion of the mountain range. Farther north, intruded masses of acidic magma solidified into rock and gave New Hampshire its nickname, the Granite State.

The retreat of the Tippecanoe Sea left the interior of Laurasia temporarily high and dry. Soon, however, the Kaskaskia Sea, a newly developed shallow-water portion of the Iapetus Ocean, flooded the barren interior. The Antler terrane, a chain of volcanic islands that presaged the emergence of the Rocky Mountains, collided with Laurasia. Acanthodian fish reached peak diversity, sharks made their appearance, and two groups of bony fish moved into the realm of freshwater: the ray-fin, which would become the dominant fish of the contemporary world, and the lobe-fin, which would give rise to four-legged amphibians that blazed the transition from water to land. The Age of Fishes was in full swim.

By Late Devonian time, an impressive variety of terrestrial habitats were in place and being overrun by big colonies of centipedes, nematodes, earthworms, leeches, and spiders. Primitive forests inhabited the nearshore wetland habitat, evolving from groundcover shrubs to canopies of leafy trees that grew to heights of 65 feet. Earth was undergoing its first green revolution. Then, approximately 360 million years ago, oceanic surface temperatures plummeted 15 degrees, and as much as 80 percent of marine species became extinct, closing the door to the Devonian period.

CARBONIFEROUS PERIOD:
359 to 299 Million Years Ago

The name says it all for the Carboniferous period. *Carboniferous* means "carbon bearing," a reference to the element found in all organic compounds. The history of this period can be condensed to one four-letter word: coal, a combustible, sedimentary rock associated with events of evolution, climate, and tectonics.

The spotlight moved from fauna to flora during this period. Mosses, ferns, scouring rushs, and horsetails clothed the warm and humid landscape of Laurasia, shaded by canopies of cycad, ginkgo, and conifer-type trees, all producing humongous masses of ground litter. In the southern hemisphere, 2-mile-thick continental glaciers waxed and waned across Gondwanaland. Average global temperatures, however, fluctuated around a moderate 54 degrees Fahrenheit, encouraging an explosive growth of plant life. All the while, Laurasia and Gondwanaland inched ever closer to one another.

The amalgamation of Laurasia and Gondwanaland deformed rock from New England south and west to Arkansas and prompted the Alleghanian orogeny, the third and final episode in the growth of the Appalachian Mountains (the Taconic and Acadian orogenies having come before). Sea-level undulations, combined with high rates of plant accumulation, enriched the eastern seaboard of the North American portion of Laurasia with more than one hundred seams of carbon-rich strata deposited within swamps and marshlands under oxygen-poor conditions. The boggy nature of these lowlands encouraged the development of both the durable, shelled amniotic egg and the plant seed, which meant both animals and plants no longer depended on an aquatic environment to reproduce. In North America, the Carboniferous period is divided into an older Mississippian and younger Pennsylvanian subsystem, based on the relative absence (Mississippian) or presence (Pennsylvanian) of coal.

PERMIAN PERIOD:
299 to 251 Million Years Ago

The annals of Earth history are replete with mass extinctions. The most significant occurred during the Permian period, the skull and crossbones chapter that ended when up to 95 percent of oceanic species became extinct, along with 70 percent of their terrestrial brethren. On an individual-to-individual basis, as many as 99.5 percent of all existing organisms died at the end of the Permian period.

The cause of this calamity is still a matter of heated debate. Did Earth collide with a celestial body, or was the culprit widespread volcanism? The setting for both theories is the same. For the first time since Rodinia had coalesced during the Proterozoic eon, a new supercontinent was forming. Shaped like the letter C, Pangaea—"one Earth"—extended from pole to pole, accentuated by three young mountain ranges: the Urals of Asia, the Alps of Europe, and the Appalachians of North America. The Tethys Sea occupied the mouth of the C, while Panthalassa—"one sea"—inundated the remaining space. Moist habitats changed to desert-type conditions, while coastal environments gave way to landmass consolidation during the formation of this supercontinent. Seed plants now dominated the forest regime, while the amphibian population began to decline, replaced by dry-air-loving reptiles. Then an event or events took place that caused the climate to change.

One likely culprit is the volume of lava, considered the largest series of lava flows Earth has ever experienced, that flooded a 130,000-square-mile-area of eastern Russia and spewed immense measures of ash and carbon dioxide, perhaps causing climate change. An opposing theory focuses on the glacial ice field of the Wilkes Land region of Antarctica, under which may lie an Ohio-sized structure that formed when a high-speed meteor impact upended mantle rock—a catastrophe that could also have brought about climate change and loss of life. Regardless of which theory is correct, climate change at the end of the Permian period came close to being fatal to all life on Earth.

For the mammals that survived this mass extinction the news was both bad and good: their antagonists, the reptiles, would soon reign supreme, but so would one of their sources of food, the cockroach.

TRIASSIC PERIOD:
251 to 199 Million Years Ago

The Triassic period, the opening chapter of the Mesozoic era, was a time of transition. The Appalachian Mountains were in full topographic bloom, the highest peaks rising to 30,000 feet. To the south the Gulf of Mexico was taking shape. Inland, eroded uplands stretched to the edge of the Great Plains and north to Greenland. Only in the west was the land submerged, beneath the shallow waters of the Panthalassa Ocean, the proto-Pacific. Across Pangaea the climate varied from warm and arid in the interior to moist and temperate in the polar and coastal regions. Continued northerly "drift" brought the supercontinent's geographic center close to the equator.

The biology of the Triassic is best described as a world transformed. Those species that had successfully crossed into the Mesozoic era enjoyed unprecedented opportunity, for vacated ecological niches abounded. Different assemblages, characterized by snails, crabs, lobsters, and new species of fish and reptiles, blanketed the ocean floor. On land, turtles, lizards, snakes, and crocodiles comingled with primitive representatives of the mammal and dinosaur communities. One would govern the Mesozoic era, the other the modern world.

The push-and-shove tectonics that had formed Pangaea gave way to stretch-and-pull forces throughout the Appalachian region, beginning a period of fragmentation. A thermal plume rising from deep within Earth's mantle stretched and fractured the undersurface of Pangaea, forming a cluster of fault-bounded troughs (grabens) from Nova Scotia to Florida. Sediment deposited in lakes impounded in these troughs are now famous rock repositories of fossilized dinosaur footprints, evidence that the arrival of the dinosaur class of reptile during this time ushered in a new era of Earth history. The Triassic ended with yet another extinction episode, but it paled in comparison to the one that had closed the Permian period.

JURASSIC PERIOD:
199 to 145 Million Years Ago

The structural destruction of the supercontinent Pangaea took place in three stages: (1) During Early and Middle Jurassic time, rifting moved from south to north, separating Africa from South America and opening a gulf that would become the Atlantic Ocean. (2) During Late Jurassic time, Africa, South America, India, Antarctica, and Australia assumed their modern identities. The Tethys Sea closed, and the Coral and Tasman seas opened up off the east coast of Australia. (3) Greenland disassociated itself from North America, and Eurasia began to assume its present-day geographic identity.

Widespread rifting built massive mid-ocean ridges that displaced tremendous volumes of seawater, causing a global rise in sea level. By Middle Jurassic time, a new inland inundation—the Sundance Sea—extended from New Mexico to Alaska, its western shore constructed of the first of many suspect terranes that would transfigure the North American landscape. These microplates, the flotsam and jetsam of supercontinent breakup, were swept up and tectonically sutured onto the leading edge of North America as it drifted to the northwest.

At the same time, the east-drifting Pacific Plate was subducted under the west-drifting North American Plate, triggering mountain building that both altered the west coast of North America and formed a volcanic island arc. The continental crust thickened above the subduction zone, and subsurface masses of granite formed the core of the Sierra Nevada and the Coast Range of western California. This entire process of mountain building along the west coast echoed the building of the Appalachian Mountains during the Paleozoic era: separate tectonic pulses that affected different regions at different times.

Two awesome species of dinosaur shared the Jurassic forests in an uneasy relationship of live and let live: stegosaurus, the 25-foot long, 2-ton, spiked plant-eater, and the long-clawed, 3-ton, meat-eating allosaurus. These Jurassic Park behemoths were a fearsome lot, but they were only a sneak preview of those to follow in the Cretaceous period.

CRETACEOUS PERIOD:
145 to 65 Million Years Ago

In word-association games, *Cretaceous* is often associated with "dinosaur," and for good reason. Miniature to humongous in size, herbivorous to carnivorous in appetite, terrestrial to aquatic to aerial in habitat, these icons of the organic world reached an astounding diversification of more than five hundred genera. During any single year probably no more than one hundred species roamed the land, but their anatomical differentiation was impressive: the 2-foot, feathered *Microraptor* was the smallest, while the 130-foot *Argentinosaurus* was the longest, and the Oklahoma-based, six-story-high *Sauroposeidon* wore the crown for being the tallest. Possibly the largest ever, *Bruhathkayosaurus* tipped the scale at up to 240 tons.

The Cretaceous saw unprecedented chalk deposition as marine waters enriched with the microscopic plates of calcareous algae extended worldwide. In North America, the 600-mile wide Western Interior Seaway extended south from the Arctic Ocean to the Gulf of Mexico, engulfing the central portion of the continent. Its western shores continually oscillated, in pulse with the second and third phases of the building of the modern Rocky Mountains.

The cause of the mass extinction that marks the end of the Cretaceous is the most contended of the five great ones that punctuate the pages of geologic history. As much as 85 percent of marine species and more than half of all land species, including the entire line of dinosaurs, became extinct, but because this happened over millions of years, cause and effect answers are difficult to find.

The key lies within a 65-million-year-old thin clay stratum —found around the world—that harbors a high concentration of iridium, an element that is uncommon within Earth's crust. Both meteoritic impact and violent volcanic eruption can create high iridium levels, fill the atmosphere with massive amounts of dust, block sunlight, and inhibit photosynthesis. Both are associated with firestorms, produce sulfuric aerosols, and emit carbon dioxide, effectively poisoning the environment. Both can eradicate life.

The necessary smoking gun for each possibility has been found. While the eruption of the thick Deccan lava flows in India 68 to 65 million years ago carries some favor within the scientific community, global scientific consensus focuses on the 65-million-year-old, 110-mile-wide Chicxulub Crater, a massive meteoritic footprint buried beneath the Yucatan coast of Mexico. Whichever the culprit, the entire dinosaur line, with the exception of birds, and many other forms of life disappeared at this time. The good news of the day was that the Mammalia class of vertebrates—the root of humanity —somehow survived.

PALEOGENE PERIOD:
65 to 23 Million Years Ago

The time line marking the beginning of the Paleogene period (composed of the Paleocene, Eocene, and Oligocene epochs), the first chapter in the Cenozoic era, is one of the sharpest demarcations in the annals of geologic history. For 100 million years two classes of vertebrates—dinosaurs and mammals—had lived side by side, but the predator-prey relationship kept one dominant and the other in the shadow of evolutionary change. The Age of Mammals now replaced the Age of Reptiles. The first mammals, rodentlike leftovers from the Mesozoic era, diversified into unfamiliar forms. For example, several species evolved from the early five-toed dawn horse, the species that many scientists believe the modern single-hoofed equine, which we are familiar with, descended from. Changes in body anatomy affected the development of many other types of mammals, including the camel, deer, cat, dog, pig, and elephant. Flights of birds disturbed the air, palms and ferns were common, the tree line extended to 80 degrees latitude, and the almost simultaneous appearance of flowering plants and pollinating insects exemplified early symbiosis.

The surface of Earth began to assume modern-day appearances. Microplates derived from an older southern landmass collided with southern Eurasia to create the Pyrenees and the Apennines in Europe and the Atlas Mountains in Africa. To the east the Himalaya Range sought altitude supremacy, the result of India colliding with Asia. In the southern hemisphere the Andean chain identified the west coast of South America.

Meanwhile, North America was undergoing geologic attack on both east and west coasts. An exploding meteor slammed into the Tidewater region of Virginia about 35 million years ago, punching a 60-mile-wide, 1-mile-deep depression in the granite basement rock and creating both the Chesapeake Bay and a tsunami wave that flushed water over the top of the Blue Ridge Mountains. In the west, the Laramide orogeny fueled a grand display of volcanism and formed a series of troughs in the Rockies that harbor shale oil reserves of more than 2 trillion barrels, many times the proven oil reserves of the Arabian Peninsula. One Wyoming trough alone contains an estimated 800 billion tons of Paleogene-age coal.

A warm, moist, and tropical climate gave way to one of the most intense warming events of Earth history. During what has been dubbed the Early Paleogene Thermal Maximum, massive amounts of carbon dioxide were injected into the atmosphere, and Arctic sea-surface temperatures rose to a steamy 73 degrees Fahrenheit—compared to some 32 degrees Fahrenheit today—causing climate fluctuations that lasted for millennia.

NEOGENE PERIOD:
23 to 2.6 Million Years Ago

The Neogene period of time—comprising the Miocene and Pliocene epochs—constitutes a mere 20.4 million years, slightly less than 0.5 percent of the 4,600 million years since the birth of the Earth. Neogene rocks are fresh, easily accessible, and relatively unaltered by the forces of weathering, erosion, and metamorphism.

Prairie, plain, steppe, and even tundra replaced the once dominant canopied forest, creating the Age of Grasses. Since grass has less nutritional value than forest plants, animal life had to adapt or die. Horses developed flat-topped teeth that continually grew upward through their gums; the flat surfaces were better for grinding the grass to get the most nutrition out of it, and the teeth were replaced as the abrasive grasses wore them down. Sea level fluctuations and volcanism produced land bridges that enabled vast herds of animals to colonize new habitats. Earth became progressively drier and cooler. Continental plate movements encouraged ice-cap formation at both the south and north polar regions. The modern world was coming into focus.

Two significant events took place during this time. The first, the Messinian Salinity Crisis, began 6 million years ago when the Mediterranean Sea underwent several cycles of complete evaporation followed by periods of refilling. The crisis ultimately concluded at the end of the Miocene epoch when the Strait of Gibraltar entry was permanently breached by the Atlantic Ocean. This event is responsible for the 2-mile-thick layer of salt and gypsum that underlies the Mediterranean Sea.

Second, tectonics created the East African Rift System. This zone of fracturing is 30 to 60 miles wide, extends a distance of 3,000 miles, and is tattooed by volcanic eruptions, earthquakes, and centers of extremely high heat. As the initial stage in the splitting of Africa, this fracture zone is a perfect analogue to the events that had begun to tear apart the supercontinent of Pangaea some 200 million years before.

QUATERNARY PERIOD:
2.6 Million Years Ago to Today

Known as both the Age of Man and the Age of Ice, the Quaternary period is the shortest of all geologic periods and is divided into two epochs: the Pleistocene and Holocene. Analyses of deep-sea cores of sediment show that the Pleistocene epoch began about 2.6 million years ago, ended 11,700 years ago, and included some thirty cycles of ice advancement and retreat. In the northern hemisphere, ice sheets extended from the Hudson Bay area south to the fortieth parallel of latitude. The last fluctuation, the Wisconsinan stage, began 110,000 years ago and eventually blanketed a 15-million-square-mile area with a thick carapace of ice—10,000 feet thick in places.

Today, U-shaped valleys gouged out by ice, the five Great Lakes, outwash plains, moraines, and abundant landscapes of fertile soil identify the Pleistocene footprint. With each advance, or buildup, of ice, the habitat of the giant ground sloth, short-faced bear, giant beaver, saber-toothed cat, woolly mammoth, and mastodon was reduced and food supply curtailed.

The final 11,700 years of the Quaternary period—the Holocene epoch—are marked by minor changes in global climate. The most recent of these, the Little Ice Age, was a period of global cooling. When it began and ended is hotly debated, but some researchers believe it began in AD 1200 and lasted for 500 years. Since the early part of the twentieth century, however, glaciers around the world appear to be in a general state of retreat, triggering warnings of the onset of global warming.

The activities of humans have now changed planet Earth enough that a new geologic epoch has been proposed. Called the Anthropocene, it is believed to have begun some 200 years ago when the human population reached one billion. Whatever the future, the geologic doctrine "the present is the key to the past" remains today as valid in the formulation of geologic history as when James Hutton first introduced its basic tenets more than two centuries ago. Ideally, knowledge of that past as it relates to the causes and effects of geologic change will inform humanity as it strives to adjust to its future.

GEO-SITES

1. Wetumpka Crater, Alabama

32° 31' 39" North, 86° 12' 17" West
Cretaceous Period Meteor Impact

This extraterrestrial invasion created the greatest natural disaster to occur in Alabama.

The majestic Appalachians dominate the geologic stage of eastern North America. Folded, faulted, and metamorphosed rocks of this range extend from Newfoundland southwesterly to Alabama, where they disappear beneath a blanket of fossiliferous sedimentary strata that was deposited in a marine setting. Usually referred to as the fall line, this junction of heavily weathered igneous and metamorphic rocks (Piedmont Province) and marine sediments (Gulf Coastal Plain) normally is not the subject of any serious geologic intrigue or inquiry.

A stark exception lies immediately east of the community of Wetumpka, where a jumble of fractured metamorphic rock belonging to the Piedmont Province breaks the otherwise uncomplicated surface. The metamorphic rock, a terrain of schist and gneiss, bends upward several hundred feet into a sharp-angled, horseshoe-shaped outcrop that is filled with a broken mélange of sandstone and shale. A small exposure of metamorphic rock reappears at the center.

In 1891, scientists mapping this unusual arrangement of rock labeled it a "structurally disturbed" area. Eighty years later, that cryptic description became cryptovolcanic explosive structure, defined as circular geometry formed by a violent release of energy and displaying local rock deformation with no direct relation to any volcanic activity. Today, it is called an astrobleme—literally speaking, a "star wound"—to indicate that the Wetumpka Crater formed when an extraterrestrial object collided with Earth, in what city officials consider the greatest natural disaster in Alabama's history.

Scientists commonly classify meteor-impact structures based on their topographic profile. A smooth, bowl-shaped depression is a simple crater. A complex crater is also a bowl-shaped depression but features a central core of uplifted and fractured rock that formed at the moment of impact, when the depressed and heated crater floor instantly rebounded in response to the dissipation of forces caused by the impact. As a textbook example of a complex crater, Wetumpka clearly exhibits three of the criteria used to establish the validity of any suspected impact:

(1) A 1998 drilling program recovered quartz-bearing rock containing planar deformation features—parallel sets of regularly spaced zones of melting caused by a hypervelocity shock wave. These form under pressures exceeding fifty thousand times normal surface conditions, and meteor impact is the only geologic process known to generate such force.

(2) Chemical analyses of the impacted rock also reveal an anomalously high level of iridium, an element uncommon in surface Earth rocks but ubiquitous to meteoritic material.

(3) Finally, the steeply dipping and circular pattern of the metamorphic rock is a typical expression of extraterrestrial impact.

Computer studies indicate a fast-traveling projectile will create a crater with a diameter approximately twenty-five times its size. Since the Wetumpka Crater is 5 miles wide, the meteor that caused it most probably was 1,000 feet in diameter. Its velocity is calculated to have been in the neighborhood of 40,000 miles per hour, and because no fragments of it have been found, it probably vaporized within a microsecond before impact.

The 1934 big band standard "Stars Fell on Alabama" memorializes the spectacular meteor shower observed across the length and breadth of Alabama on November 12, 1833. The lyricist could not have known that 83 million years before—a time determined by the age of the youngest disturbed rock within the crater—one gargantuan star catastrophically crashed into the shallow sea then covering the southwestern portion of the state, creating what is today locally lauded as the best-preserved marine impact crater in the world.

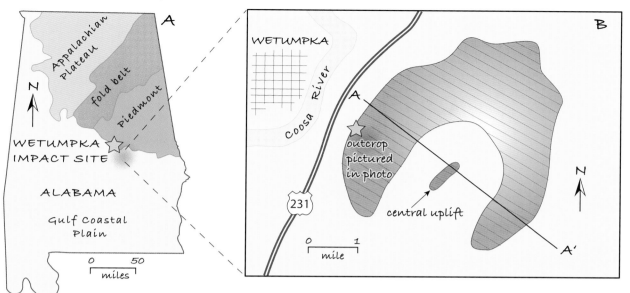

(A) The Wetumpka Crater lies near the contact of two geologic provinces: the Piedmont of the Appalachian Mountains and the Gulf Coastal Plain. (B) The zone of disturbed rock (orange) covers a 25-square-mile area.

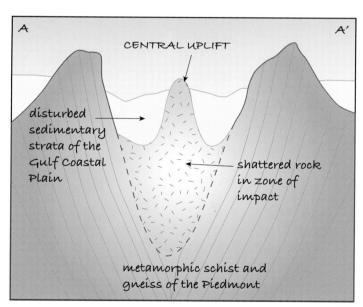

The central uplift of shattered rock in the Wetumpka Crater is evidence of a complex crater.

Because the eastern suburbs of the city of Wetumpka are built over the impact site, the steeply dipping rock exposed behind the building complex where Breezehill Boulevard intersects US 231 is the best place to observe this feature. The rock was bent this way due to the force of the meteor impact.

2. Exit Glacier, Alaska

60° 10' 52" North, 149° 38' 19" West
Pleistocene Epoch Ice in Motion

Become acquainted with a living glacier up close and personal.

Over the past 2.6 million years, as planet Earth received an ice-sculpted facelift courtesy of the ebb and flow of Quaternary-age glaciers, many an undistinguished terrain became not merely topographically unique but geologically eccentric. Without glaciers, some of the great physiographic icons of the world would not exist: the Great Lakes, the Matterhorn on the Italian-Swiss frontier, the fjords of Norway, and the Yosemite Valley of California. At the height of the last ice age, around 12,000 years ago, frozen water weighted down approximately 30 percent of Earth's land surface. Today, that figure is listed as 10 percent and dropping, with debate increasingly focused on the "dropping" part: is it part of nature's climatic swings, or does it come from human activity?

A glacier is a large mass of compacted and recrystallized snow that flows down or outward in all directions because of gravity and the pressure of its own weight. While scientists still debate the exact causes of glaciation—the widespread development of glacial ice—three necessary landmass conditions supposedly set the stage: a geographic position close to the poles, high elevation, and the abundance of atmospheric moisture that comes with latitude and elevation. Today across the globe, age-old ice packs are generally in retreat—some melting a few inches each year, but others at a pace more like a popsicle on a summer sidewalk. Even Glacier National Park, that bastion of "cool," may soon become just plain National Park, since only 26 of the 150 glaciers catalogued there in 1850 remain.

As in any review of dire geologic circumstances, exemplary examples exist. In Alaska, an estimated 100,000 glaciers cover some 5 percent of the land surface—almost 30,000 square miles. Most are isolated and remote, but Exit Glacier has earned repute because of its beauty and park-and-walk accessibility. This ice-blue crown jewel of Kenai Fjords National Park rises majestically from the landscape some 12 miles northwest of Seward, Alaska.

Early on in the Quaternary period, the Kenai Mountains of coastal Alaska became the gathering ground of snowfall that exceeded today's not unusual rate of 80 feet per annum. With accumulations year after year exceeding seasonal melt, snow evolved to ice and eventually formed the mountain-encompassing Harding Ice Field, which, at 5,000 feet thick and home to at least thirty-five glaciers, is the largest of its type extant in North America. Exit Glacier alone measures 3 miles long and 0.5 mile wide and cascades 2,500 feet through a tributary to the Resurrection River, changing what had been a V-shaped course into a characteristic, ice-carved, U-shaped profile.

This tongue of fractured and crevassed ice is both growing and dying. Each day it advances as much as 2 feet, propelled by internal deformation, by slipping at its base, and by fracturing of its brittle portions. The snap, crackle, and pop of tortured ice forms the theme music. Simultaneously, however, along with 90 percent of Alaska's glaciers, Exit is dying. Season by season, the total mass is diminishing, presumably because of global warming. It once extended all the way to Seward.

Walkways extend to the very edge of Exit Glacier, and then upward to the parent Harding Ice Field. Nunataks, glacier-eroded basins, lateral and terminal moraines, ice caves, and yawning crevasses all portray the awesome prowess of ice in motion, where one step seems like a tour of the entire Pleistocene epoch.

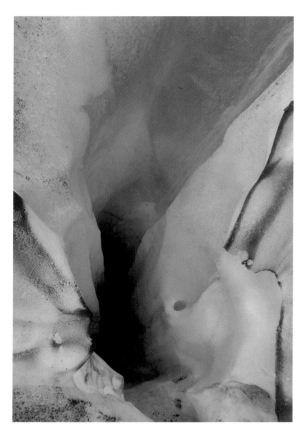

Blue-ice moulins, scoured by descending eddies of meltwater, punctuate Exit's surface.
—Courtesy of Fiona Ritter-Davis, National Park Service

An outwash plain of dark sand and gravel accentuates the present-day front of Exit Glacier. —Courtesy of Fiona Ritter-Davis, National Park Service

Numerous exposures of bedrock (nunataks) pierce the crown of the Harding Ice Field. —Courtesy of Fiona Ritter-Davis, National Park Service

3. Antelope Canyon, Arizona

36° 50' 57" North, 111° 22' 10" West
Neogene Period River Erosion

A marvel of form and color created by the ever-evolving interplay of fractures and flowing water.

The Canyonlands physiographic province of the American Southwest is a visual mélange of escarpment, plateau, mesa, and canyon. This is classic red rock country, where the landscape is at once unashamedly naked, devoid of all but the occasional desert plant, and unpretentiously clothed in an infinite variety of earth tones. The forces of geology have magically conducted their affairs here for millions of years, carving, polishing, and buffing the land to a state of near artistic perfection. In every sense of the word, this is *awesome* country.

Beneath this panorama of beauty and form lies another world of wonders generally unseen by the casual traveler. This is the realm of the slot canyon, a narrow, undulating, bare rock crevice that deepens into a semiclaustrophobic, sheer-walled foyer periodically bathed in incandescence.

Slot canyons are found throughout the world, created by the proper combination of fractured rock, preferably sandstone or limestone, and intermittent rainfall. Perhaps the highest concentration—more than one thousand—dots the high plateau of southeast Utah and northeast Arizona. Most are remote and inaccessible. Antelope Canyon, one of the rare exceptions, lies in the Navajo Indian Reservation east of Page, off mile marker 299 on Arizona 98. Here a combination of ambient light, rock color, depth, and width creates a visual effect that earns the title of the most photogenic slot in the world.

Antelope Canyon is divided topographically into two separate sections: lower and upper. Most visitors choose the upper section, entering through an unassuming and irregular crack in a column of oxidized rock that opens quickly into a surreal chamber, 120 feet high, with sinuous walls of crossbedded sandstone. Photographers prefer mid-April to early September, when the sun is high in the sky and cascading rays of light enhance the already mesmerizing palette of color.

Over time, floodwaters flowing through Antelope Creek, which empties into nearby Lake Powell, carved the canyon out of the underlying bedrock of Navajo sandstone, a sedimentary formation that outcrops across several southwestern states, most notably at Canyonlands, Capitol Reef, and Zion national parks. During the early part of the Jurassic period, prevailing winds deposited sand dunes along the coast of a sweeping inland sea that inundated North America. As the sand dunes grew and migrated, foot-scale crossbedding developed as layer upon layer of sand was deposited. The sand dunes were later fossilized, becoming the Navajo sandstone, and minute amounts of iron oxidized to create the kaleidoscopic range of salmon, pink, vermilion, orange, and crimson colors. Iron is much like food coloring—a little dab is all that is needed.

Both sections of Antelope Canyon began to form when the elements sought out a key fracture within the massive architecture of the Navajo sandstone, eroding and widening the fracture. Flash-flood waters infiltrated the narrow passageway, carrying loads of sand and silt that abraded and deepened it. Surging water plucked out rock fragments as it smoothed and polished the walls. Over time the slot grew in depth and grandeur.

Slot canyons can be dangerous. In 1997, in the lower section of Antelope Canyon, rapidly rising water generated by a storm 5 miles upstream took the lives of eleven people. Because of this tragedy, guides authorized by the Navajo Nation now accompany all tours. As in many aspects of nature, beauty and disaster have learned to lurk here in tenuous coexistence.

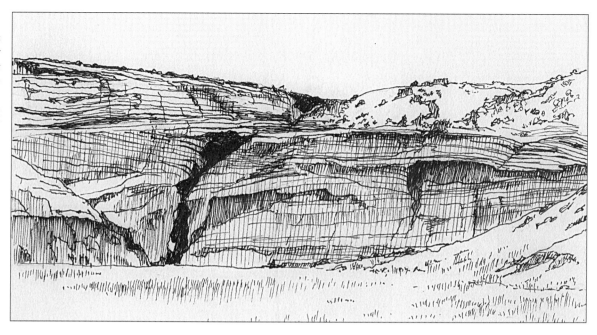

The unpretentious portal of the upper canyon gives little hint of the fantasyland that lies within.
—Drawing by Patsy Faires, Kernersville, NC

Curvilinear walls of this slot are awash with a three-dimensional swirl of color that continually evolves with changes in light intensity and refraction. —Courtesy of Charles Herron, Wilmington, NC

The collage of ruffles, protuberances, and contorted surfaces was created as water scoured the slot.
—Courtesy of Charles Herron, Wilmington, NC

4. Meteor Crater, Arizona

35° 01' 37" North, 111° 01' 21" West
Pleistocene Epoch Meteor Impact

The decoding of these desert rocks explained the what, why, and how of worldwide impact sites.

Large, circular structures pockmark igneous, metamorphic, and sedimentary rock terrains around the world. For decades scholars heatedly debated their origin. Was cryptovolcanic process—an internal, concealed, gaseous, volcanic-style of explosion—the cause? Or did a bolide, a large, exploding body, slam into Earth? The debate was one strictly of pure science until America's plan to travel to the moon made it imperative that the many circular features disrupting its surface be fully understood—a case of applied science. As evidence in support of the hypervelocity impact of bolides mounted, cryptovolcanic advocates began to shift their support to that hypothesis.

South of I-40 and halfway between Holbrook and Flagstaff, there is a 600-foot-deep, 0.75-mile-wide, circular astrobleme—"star wound"—in the Colorado Plateau. Called Meteor Crater, it appears to have formed just yesterday, rendering it a natural site for field study.

Circularity is essential, but not sufficient to establish whether or not a crater in Earth's crust is a meteor impact site. Proof requires at least one of five other features:

(1) Either coesite or stishovite particles are present. The common mineral quartz has a density of 2.66 grams per cubic centimeter (g/cm^3). Impact forces more than 25,000 times greater than normal atmospheric pressure will metamorphose quartz to coesite, with a density of 2.93 g/cm^3. Ratchet the pressure up to 130,000 times normal and the result is stishovite, with a density of 4.35 g/cm^3.

(2) Planar deformation features are created by impact forces of at least 50,000 times normal atmospheric pressure. Under a microscope, they appear as multiple parallel sets of regularly spaced zones of fracture within grains of quartz.

(3) Shatter cones, striated, dunce-cap-shaped features, also form within rock subjected to elevated pressures. They occur in nested groups, with the tips of the cones pointing toward the center of the impact site.

(4) The presence of meteorite fragments is a dead giveaway.

(5) Finally, unless an impact site is subjected to extensive erosion, an encircling deposit of ejecta—masses of angular and broken rock formed at the instant of impact—generally marks the spot.

At Meteor Crater both coesite and stishovite, as well as more than 15 tons of meteorite fragments, have been collected around the 150-foot-high ejecta rim. Impact at this site is thus fact, but how it happened is the topic of multiple computer-generated scenarios.

A typical version posits that a meteor, dubbed the Canyon Diablo Meteor, approached Earth measuring 130 feet in diameter, weighing 300,000 tons, and moving at more than 27,000 miles per hour. The force of its impact, equivalent to the detonation of 20 million tons of TNT, created a hemispherical shock wave that coursed down and outward, gouging out the crater. Ejecta debris splashed over 100 square miles, and an all-consuming fireball raced outward at 1,200 miles per hour. Thermoluminescense analyses—a laboratory process useful in the age dating of rock metamorphosed by high-energy impact—reveals the meteor struck Earth some 49,000 years ago.

Of the more than 180 impact sites recognized around the world, approximately 25 are in the United States. The recognized godfather of them all, Meteor Crater is a perfect example of a simple crater—one without a central uplift (see geo-site 1). It is the most thoroughly investigated and by far the best preserved of the many astroblemes that scar the face of planet Earth.

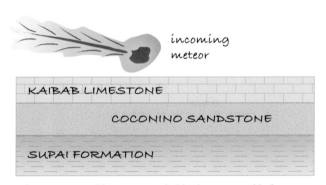
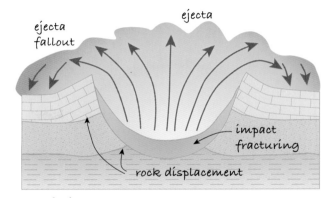

The Canyon Diablo Meteor probably disintegrated before impact, creating a shock wave that formed a crater many times its size. Fractured rock covered its floor.

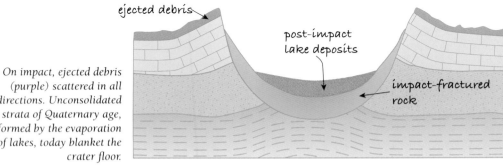

On impact, ejected debris (purple) scattered in all directions. Unconsolidated strata of Quaternary age, formed by the evaporation of lakes, today blanket the crater floor.

Meteor Crater served as an outdoor laboratory for American astronauts preparing to examine similar sites on the moon.
—Courtesy of Meteor Crater Enterprises, Flagstaff, AZ

5. Monument Valley, Arizona

36° 58′ 58″ North, 110° 06′ 44″ West
Cenozoic Era Desert Erosion

A wonderland of mesas and buttes created over the expanse of years by everyday forces of erosion.

In the history of Earth, as in the daily press, calamitous news often forms the headlines. Events of catastrophic consequence punctuate the chapters that constitute the continuum of geologic time. For example, the great mass extinction 251 million years ago of up to 95 percent of all oceanic species, including the iconic ancestral crustacean, the trilobite, was reason enough to say good-bye to the Paleozoic era.

Nonetheless, any pragmatic Earth history worth its salt should occasionally mix the shocking and disturbing with a chapter lacking tragedy and violence. The story of northeastern Arizona's Monument Valley—one of the most famous and beautiful landscapes in the world—is one such "lacking" chapter. Here, picturesque, isolated remnants of a former landscape speak of a peaceful sedimentary birth, a youth and middle age distinguished by uncomplicated uplift and stability, and now an undisturbed old-age slide toward death by erosion.

The rocks of Monument Valley are principally identified with the Organ Rock and DeChelly formations, deposited during the final years of the Permian period, 270 to 251 million years ago. The environment they were deposited in ranged from coastal lowlands drained by sluggish rivers to plains of windblown sand much like today's Sahara Desert. For millions of years these regional, near-sea-level conditions remained unaltered.

Change came, however, with the onset of the Laramide orogeny, the final pulse of compressional forces that folded and faulted the rock framework of western North America. Add localized metamorphism and inundation with floods of lava, and the eventual result was the modern Rocky Mountains. Amidst this sea of tectonic torture and disarray, the Colorado Plateau and its bull's-eye center, the ancestral Monument Valley, stood out in a state of simplicity. Here, a 130,000-square-mile chunk of crust kept its original horizontal orientation even as it rose 7,000 feet like an oversized concert stage elevator.

For the past 65 million years the undulating meanders and expanding floodplains of the tributaries of the San Juan and Colorado rivers have eroded this high semidesert, sculpting it into an evolving catalogue of flat-topped landforms—mesa, butte, and pinnacle—that boldly accentuate the horizon. The size of the summit areas of these plateau sentinels defines them—mesa the largest, and pinnacle the smallest. Rock character determines their topography. Organ Rock shale weathers to a sloping profile; hard DeChelly sandstone, to vertical cliffs. The Shinarump Member of the Chinle Formation of Triassic age, a very hard, erosion-resistant conglomerate, typically forms a protective caprock to these landforms.

Today, frost action, running water, and wind are the principal forces of erosion. As freezing water expands cracks and fissures, blocks of rock break off and weather to sand-sized material that prevailing winds and moving water then transport to base level, the lowest elevation toward which the forces of erosion can operate. While temporary base levels exist locally, the ultimate base level for Monument Valley is sea level.

When the final pinnacle of Monument Valley disappears, a chapter of consequence will have been written. This is a peaceful story of how a low-lying coastal plain escaped catastrophic episodes of geologic upheaval to gradually evolve to a towering plateau, only to be reduced back to sea level by the ageless assault of wind and water. Throughout the often-devastating annals of geologic history this vignette stands as a most unusual cycle of unheralded tranquility.

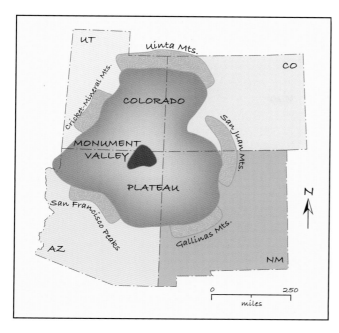

Monument Valley lies at the center of the Colorado Plateau, a core of undeformed rock enveloped by mountainous terrain.

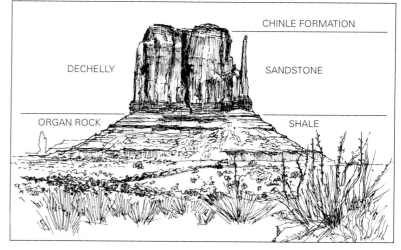

Weathering and erosion along vertical fractures created the thumb of West Mitten Butte. —Drawing by Patsy Faires, Kernersville, NC

This semi-silhouetted view of West Mitten Butte (left) and its mirror image, East Mitten Butte (right), is one of the most photographed vistas of North America.
—Courtesy of Charles Herron, Wilmington, NC

6. Prairie Creek Pipe, Arkansas

34° 01′ 55″ North, 93° 40′ 22″ West
Cretaceous Period Magma Intrusion

The singular location in the world where folks can keep any and all diamonds that they find.

A naturally occurring crystalline form of carbon, diamond is the hardest of the approximately four thousand known minerals. Its outstanding ability to separate light rays into their color components produces unsurpassed brilliance and "fire," and diamonds have been traditionally associated with concepts of wealth, status, and everlasting love.

Because of their high density, 3.52 grams per cubic centimeter, diamonds often accumulate in river placer deposits, where gravity segregates the heavier, diamond-bearing weathered rock debris into gravel deposits. Placer deposits can easily be mined by either pick and shovel or high-pressure, hydraulic means. In contrast, commercial operations involve heavy-machinery extraction of diamond-bearing material from some of the largest open-pit mines in the world.

Diamonds are often associated with lamproite pipes, vertical bodies of commonly mineralized igneous rock. More than often they are shaped like long-stemmed martini glasses. Lamproite pipes generally have a diameter of 2,000 to 3,000 feet and plunge to depths of hundreds of miles. Diamonds are speckled throughout the lamproite mass.

One hundred million years ago, an eruption charged by carbon dioxide and steam violently disrupted the Early Cretaceous–age bedrock of the Crater of Diamonds State Park, 2 miles southeast of Murfreesboro on Arkansas 301. Before the eruption, intruding magma had cooled 100 miles beneath the surface in the mantle, forming an igneous body of rock containing diamonds. Then explosive volcanic activity shattered the igneous body and surrounding rock, flushing the diamond-bearing material toward the surface. The explosion formed the funnel-shaped crater that the ejected material, including that with the diamonds, fell back into. The deposits in the pipe include ash and lapilli (small, solidified fragments of lava), breccia (angular fragments of rock encased in a finer-grained matrix), and lamproite (mantle-derived volcanic rock with a high potassium and magnesium content). The Prairie Creek pipe was born. Exploratory drilling indicates the zone of breccia, that section open to the public on a what-you-can-find-you-can-keep fee basis, is 670 feet deep and has a volume of 78 million tons of probable diamond-bearing rock.

This pipe is the largest of seven sister pipes in the Prairie Creek region, which is part of the Ouachita-Appalachian Mountain Front, a curvilinear diamond-discovery belt that extends from Arkansas through Alabama, Georgia, North and South Carolina, and into Virginia. This front formed during the late Paleozoic era as part of the assembly of the supercontinent Pangaea.

Of the roughly five thousand potential diamond-bearing structures in the world, only an estimated 1 percent are rich enough to be commercially mined for diamonds. Between 1972 and 2009, 28,745 diamonds were found in the Prairie Creek pipe. One individual found a total of 85 during a twenty-seven-year period. While 834 of these finds register over 1 carat—3 percent of all those found—most are the size of a match head. The most common diamond colors found at the pipe are white (60 percent), followed by brown and yellow. The American Gem Society grades the Strawn-Wagner 3.03-carat specimen, discovered in 1990, as 0/0/0, meaning it is perfect in cut, color, and clarity, a rare gemstone designation. The largest find to date, and the largest ever discovered in the United States, is the Uncle Sam diamond, weighing 40.23 carats.

An estimated three hundred thousand gems are yet to be found in what is considered the eighth-largest diamond reserve in the world. At an average discovery rate of seven hundred diamonds per year, a statistic maintained over the past three decades, more than four hundred years of successful explorations are yet to take place.

Coarse-grained breccia forms the southeastern portion of the Prairie Creek pipe exposure. Ash, lapilli, and lamproite comprise the remaining area. The breccia contains 0.01 to 1.25 carats of diamonds per 100 tons of rock. The rest of the pipe is barren.

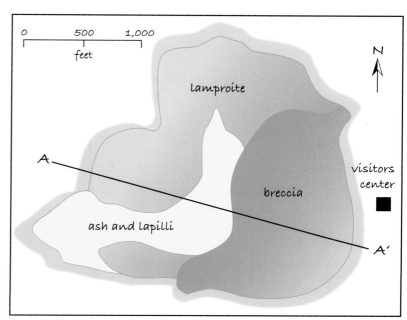

Recovered diamonds include Star of Shreveport (8.21 carats), Black Beauty (1.45 carats), and the flawless yellow Sunshine diamond (5.47 carats). —Courtesy of Crater of Diamonds State Park

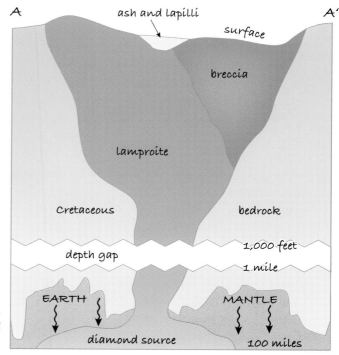

Cross section showing the relationship of the exposed rock and deeply buried diamond source.

7. Wallace Creek, California

35° 16' 16" North, 119° 49' 36" West
Quaternary Period Earthquake Movement

A visual example of catastrophic topographic disruption along a world-famous fault.

The eight or more continental-sized tectonic plates that comprise the crust of the Earth function like the segments of a turtle's shell, forming the surface and protecting the interior. In constant motion, they crush together, move apart, and grind past each other, forming major faults along which movement takes place during an earthquake. The San Andreas Fault in California, traceable at the surface for 800 miles from Cape Mendocino to the Salton Sea, is known as a transform fault, since it is the contact between two major tectonic plates that are grinding laterally against each other. Because of its reputation as a cataclysmic and destructive killer, and the ease with which it can be studied, this is probably the most analyzed transform fault in the world.

Nowhere is the detail of San Andreas displacement better illustrated than where Wallace Creek exits the Temblor Range and begins its passage across the Carrizo Plain about 16 miles west of McKittrick. From California 58 turn left onto Seven Mile Road and follow it for 0.3 mile. Turn left again onto Elkhorn Road and follow it for 4 miles to the Wallace Creek parking lot.

This geo-site straddles the center of a 220-mile segment of the San Andreas Fault that ruptured most recently on the morning of January 9, 1857, with a Richter scale magnitude of 7.9, a release of energy equivalent to the explosion of 800 million tons of TNT. Within less than two minutes, it displaced every ridge, gully, and tree line in the region 30 feet in a right-lateral direction.

At Wallace Creek, the Pacific Plate (Carrizo Plain side) is moving to the northwest relative to the North American Plate (Temblor Range side). This displacement takes place at an average pace of 1.3 inches per year, similar to the rate at which the human fingernail grows. The movement, however, is not continual. Many—indeed, most—faults do not experience significant motion for tens or even thousands of years. Large, infrequent energy releases interspersed with a multitude of minor earthquakes are more the norm. Studies have shown the 1857 quake occurred along a segment of the San Andreas that experiences a major rupture on average every 130 years, which means this area is theoretically overdue for an earthquake of significant magnitude.

The present-day channel of Wallace Creek is offset 420 feet where it crosses the fault. Carbon-14 age dating of charcoal samples collected from soil layers of the creek indicates it formed 3,800 years ago. At that time, it cut a drainage course straight across the San Andreas Fault. Over the passage of many centuries, motion due to infrequent earthquakes has caused the upstream portion to shift to the southeast. Information such as this makes the Wallace Creek site of paramount importance in the better understanding of the San Andreas Fault and its propensity to destroy property and take lives.

Assuming the average displacement of 1.3 inches annually continues, and there is no reason to think it will abate, today's metropolis of Los Angeles will snuggle up comfortably against present-day San Francisco in 17 million years, a mere pittance of geologic time. By then, hopefully all the secrets of the Wallace Creek portion of the San Andreas Fault will have been unearthed, to the betterment of society—if one still exists.

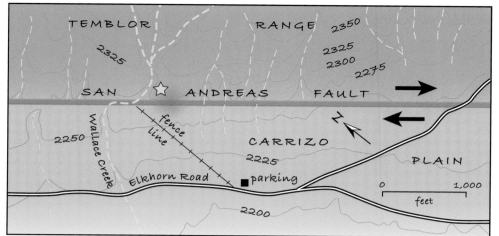

Wallace Creek flows through the Temblor Range, crosses the San Andreas Fault, and continues to the southwest over the Carrizo Plain. The star identifies the view site. Elevations are in feet. Dashed lines represent intermittent streamflow. (Modified after Sieh and Wallace, 1987.)

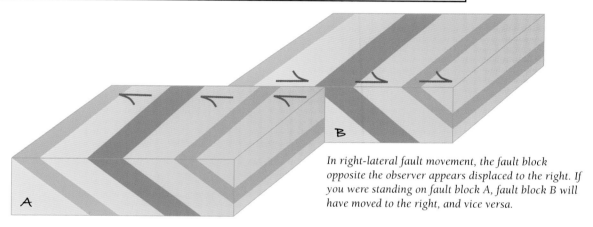

In right-lateral fault movement, the fault block opposite the observer appears displaced to the right. If you were standing on fault block A, fault block B will have moved to the right, and vice versa.

The 420-foot offset of Wallace Creek (dashed line) is perhaps the most photographed feature of the San Andreas Fault. The view here is to the northwest. Arrows denote direction of relative movement along the fault.

8. Racetrack Playa, California

36° 40′ 08″ North, 117° 33′ 22″ West
Holocene Epoch Wind Erosion

No one has ever seen them move, yet these rocks continue to dance across the floor of Death Valley.

More than one hundred years ago, rumors began to seep out of the goldfields of California. A prospector had discovered boulders weighing up to 700 pounds haphazardly scattered across a high-altitude desert floor. Behind the rocks stretched grooved trails—some straight, some sinuous, some doubling back upon themselves, and some 3,000 feet long. No one has observed the obvious motion implied by the trails, but geologists have glibly described the phantom movement as racing, sailing, sliding, roving, scuttling, traveling, wandering, and even dancing.

Racetrack Playa is nestled 3,708 feet up between the Last Chance Range and the Cottonwood Mountains of California, in the remote northwestern sector of Death Valley National Park. This smooth, vegetation-free, former lakebed is 1 mile wide, stretches 3 miles through the desert, and is dampened by a mere 2 to 4 inches of rain per year.

At the southern end of the playa, rock trails are mainly linear and point to the northeast. Farther north, they are more sinuous. All are shallow and have smooth, rounded, and slightly elevated ledges that suggest a plowing motion. Details of the furrows indicate the rock movement is episodic and that the rocks rotate as much as 150 degrees. Composed of dolomite, a magnesium-rich limestone, the boulders with rough bases tend to produce the straightest paths, while those with smooth bottoms create curved routes.

Several theories have been developed to explain this phenomenon. Once considered the culprit, gravity was dismissed when surveys revealed that the playa floor rises a mere 2 inches to the north, the direction the typical rock trace takes. Gravity, after all, moves things downslope, not upslope.

How about magnetism? This theory also lacks credibility because magnetic minerals are completely absent in these Cambrian-age rocks of mystery, as well as in their obvious source, a nearby 800-foot-high slope. No magnetic minerals, no magnetism!

Perhaps gale-force winds move the rocks after the lake floor is greased by rainfall. Reported on occasion to blow out automobile windows and force hikers to their knees in Death Valley, wind enters the playa through two topographic channels. The principal direction is to the northeast, parallel to the longest and straightest trails. The other is northwesterly, supposedly the cause of the more sinuous furrows.

Given, then, a combination of wind and water, could precipitation of any kind, be it fog, dew, rain, frost, or snow, form an extremely slippery surface—in effect reducing the coefficient of friction—and create conditions under which strong air currents move selective rocks? Or, since many of the trails are parallel, could ice crystallize from a thin layer of water in the lakebed and temporarily lock the rocks in an ice harness? Wind blowing against the surfaces of the imbedded rocks might then break the frozen sheet into ice-raft sections, each of which would then move in stages, causing the entrapped rocks to leave parallel traces in the wet clay surface of the lakebed. Eventually these rafts would melt, freeing the rocks and terminating the tracks.

Theories abound, but consensus is absent to explain the lithic travelers of Racetrack Playa. Scientists still ponder rock shape, mass, and angularity; the duration and rate of precipitation; the extent and depth of water saturation in the playa floor; and the percentage of clay sediment and the presence or absence of gelatinous algae, both of which could act as lubricants. The dance isn't yet over and the beat goes on, but at least the melody and rhythm are better understood.

At this geo-site of mystery, traveling rocks derived from the hills in the background glide across a mud-cracked dance floor.
—Courtesy of Jay D. Snow, National Park Service

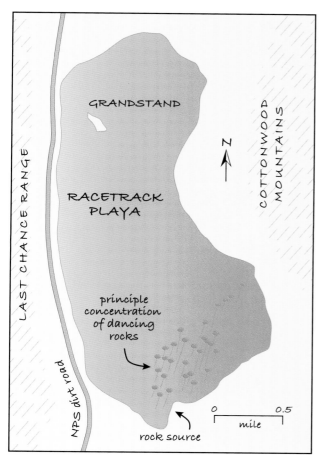

The parking lot at the southern end of Racetrack Playa offers the best access to this dancing rocks arena.

Rock-crowded portions of the Racetrack Playa floor produce curious intersecting sets of curvilinear skid marks. —Courtesy of Jay D. Snow, National Park Service

35

9. Devils Postpile, California

37° 37' 31" North, 119° 05' 06" West
Pleistocene Epoch Magma Intrusion

An imposing wall of stacked columns with a fire and ice heritage.

The last great age of glaciation began 2.6 million years ago, when an icy cloak settled across the northern and southern hemispheres. While the presence of glaciers in Antarctica and Greenland indicates this climatic act continues in diminished effect, scientific consensus supports the idea that spaceship Earth is presently moving through an interglacial epoch, a period of relative warming when glaciers recede toward the poles, and will in the near geologic future again enter a period during which glaciers expand across the continents. Even when glaciers were widespread during this most recent ice age, there were episodes of high-temperature volcanic activity. The igneous wonder of Devils Postpile National Monument, located 14 miles southwest of the community of Mammoth Lakes, is a prime example.

Less than 100,000 years ago, a very liquid form of magma poured onto the mountainous surface of eastern California and coursed its way southward along a glaciated drainage that would millennia later be labeled the Middle Fork San Joaquin River. When the molten rock ran into an obstruction, perhaps a ridge of unsorted debris left by a retreating mountain glacier, it pooled to an estimated depth of 400 feet and began the slow process of crystallization. As it cooled to ambient temperature, it formed an imposing structure that today is considered one of the best examples of columnar basalt on Earth.

Compared to other exposures of columnar basalt across the globe, Devils Postpile is world-class because its columns are perfectly developed and elegantly exposed. They average 2 feet in diameter and extend to heights of more than 60 feet. More than 90 percent are five- or six-sided; the rest, three-, four-, or seven-sided. Stacked like tall telephone poles planted in a talus of broken "posts," many are broken into sections that rest precariously on fractured bases. Earthquakes occasionally cause a segment to tumble down.

Because the top of the postpile lava flow was exposed to cool mountain air and its base rested on cold bedrock, it cooled inward from both surfaces. As it solidified, tensional stresses caused the rock to crack, following the rule that all liquids, except water, occupy less space when "frozen." In a mass of lava that cools uniformly, is homogeneous in composition, and is of uniform thickness, the developing fractures grow laterally from the cooling surfaces until they join one another to produce hexagonal columns. Because these ideal conditions are seldom achieved, the result is often irregularly shaped, polygonal columns bounded by curved fractures and a variable number of sides. If the cooling surface is not level, the columns may even lie at angles away from the vertical.

The top of the postpile is a must-photograph mosaic of interlocked, multisided column ends exposed like a tile floor that has been scoured and polished. The thousand-foot-thick ice age glacier that scraped off the crown of the Devils Postpile—grinding away with its abrasive base of rock fragments, sand, and silt—completed a saga that began with volcanism and evolved through glaciation. This monument, contained within a mere 798 acres, is small in size but immense in value. It is an outstanding example of how Mother Nature can combine the contrasting effects of fire and ice to sculpt a minilandscape exceptional within the American geologic scene.

Two classes of columns, one long and linear, the other twisted and contorted, compose Devils Postpile. Periods of freezing and thawing have broken both sets into segments that form the pile in the foreground.

This fallen segment, measuring 18 by 30 inches, defies geometric description. Does it have four, five, or six sides? Pen for scale.

Striations scratched into the eroded columns by sediment encased in ice indicate which direction the glacier moved—parallel to the scratches. Darker areas highlight what remains of the glacial polish. Pen for scale.

10. Rancho La Brea, California

34° 03' 47" North, 118° 21' 21" West
Pleistocene Epoch Fossilization

Intriguing secrets are revealed in this almost 50,000-year-old asphalt graveyard of ice age animals and plants.

The twenty-first century ushered in multiple crises on the world scene, not the least of them the realization that the era of cheap oil is about to end. Do the rocks of planet Earth contain a finite volume of this fossil fuel? Most certainly! Has most of it been consumed by the engines of progress? Certainly not! Since the Age of Oil began some 150 years ago, approximately half of the world's reserves have been consumed. This was the easy-to-find half. Discovering new reserves will increasingly require ever more costly and sophisticated exploratory technologies.

Early on the process of discovering oil in the field was simple: drill where there was an oil seep—the surface evidence of a permeable pathway to a subterranean reservoir of petroleum. Edwin Drake sited his 1859 Pennsylvania borehole, the first drilled specifically for oil in the United States, near seeps used by fifteenth-century Native Americans (see geo-site 69). Between the 1860s and the early 1900s, geologists discovered new fields in California wherever they found a seep.

Of all the known seep locations in the United States, one is unique: the Rancho La Brea Tar Pits in Hancock Park, along the 5800 block of Wilshire Boulevard in downtown Los Angeles. For almost 50,000 years this 23-acre site has oozed crude oil from depths of 1,000 to 6,000 feet, creating one of the richest, best-preserved, and most intensely studied collections of ice age fossils in the world. Paleontologists have recovered the remains of 60 species of mammals (including ground sloth, dire wolf, short-faced bear, camel, mammoth, and saber-toothed cat), 135 species of birds (such as California condor, eagle, and falcon), and numerous plants and insects—together totaling more than 620 species.

Throughout the latter stages of the ice age, and continuing to the present, flows of petroleum have leaked onto the surface and begun the slow transformation to asphalt, first by the evaporation of methane gas and then by chemical alteration by bacteria. This oil came from mudstone and shale strata belonging to the Miocene-age Monterey Formation, a rock rich in the fossilized shells of diatoms. Long ago the influences of temperature, pressure, and time had changed the organic content of these microscopic, single-celled marine plants to petroleum.

During hot summer months many an animal wandering upon this site was entrapped within the gooey, viscous asphalt, their cries of panic quickly attracting multiple predators intent on an easy meal. Many of them, too, became victims. While this scenario explains the known ratio of seven carnivore skulls to the remains of every single herbivore, the uneven age-date distribution among the thousands of extracted La Brea bones suggests an animal was probably entombed only once every decade or so. After the flesh rotted, the pores of the bones became saturated with asphalt, a decay-inhibiting preservative used as an embalming agent since the days of the Egyptian pharaohs.

The oldest La Brea fossil dates to 46,800 years ago. After the dire wolf, the most common is the saber-toothed cat, the infamous and extinct, 6-foot-long, 750-pound, pouncing predator of the ice age. Surprisingly, the remains of only one human are known—a partial, 9,000-year-old skeleton of a female. Exploration continues. Today, at the on-site Page Museum, visitors can observe scientists cleaning fossils recently extracted from the tar pits.

Iridescent bubbles of methane gas percolating in a sea of petroleum froth show chemical activity is alive and well at Rancho La Brea. —Courtesy of Sarah Lansing, Pasadena, CA

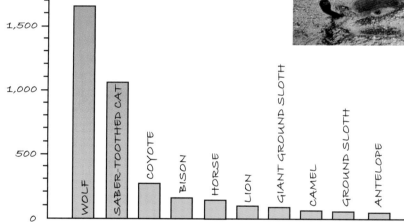

Most La Brea bones are those of carnivores (wolf, saber-toothed cat, coyote, and lion), the rest, herbivores. This ratio of hunter to hunted is chilling evidence that this was a popular killing ground.

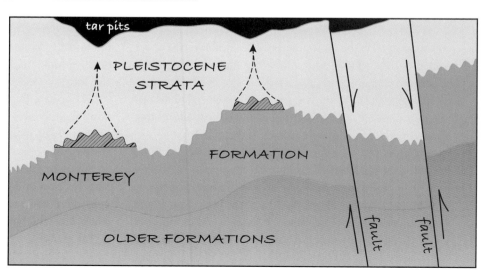

The La Brea Tar Pits formed through the upward migration and chemical alteration of petroleum deposits trapped in the Monterey Formation.

11. El Capitan, California

37° 44' 02" North, 119° 38' 13" West
Neogene Period Ice and Water Erosion

This granite monolith is the crown jewel of a terrain sculpted by water, ice, and time.

The topographic spine of California is high country, sustained majestically by the Sierra Nevada, a mélange of Mesozoic-age granitic batholiths that weld the southern city of Barstow to the northern community of Susanville. Home to the gold rush of 1849 and Mount Whitney—at 14,505 feet the highest point in the contiguous United States—the mountains are also the location of Yosemite National Park, where the photography-inspiring arena of alpine vistas is a model for studying how ice and water erosion affect an igneous terrain. The park's Rhode Island–sized enclosure is chock-full of sculpted iconic features, among them El Capitan, at 7,569 feet high the largest solid mass of granite in the world. This looming monolith is a commanding presence beside the meandering Merced River in the center of the 0.5-mile-deep, U-shaped Yosemite Valley.

For well over a century, El Capitan has attracted two kinds of people: those seeking heights, and those seeking understanding. It was not until 1958 that cliff-face climbers made it to the top, after forty-seven days of painstaking progress up and over the Nose, the prow that divides the sheer wall into two cheeks of rock labeled simply "southwest" and "southeast." The other quest involves the celebrated geomorphology of the park. Is it the result of glacial abrasion? River erosion? Or does its heritage relate to the fractures that transect many of the granite masses of the park?

El Cap is one of approximately fifty deep-seated igneous intrusions (plutons) that invaded the depths of Yosemite 135 to 80 million years ago. It is composed principally of pale, coarse-grained granite, the same rock type found within most of the landforms in the western end of Yosemite Valley. A separate intrusion of dark rock bisects the vertical wall with a suggestive outline, thus its designation "map of North America." To the left a dike diagonally scars the nose of El Cap.

As erosion exhumed the plutons of Yosemite National Park from their emplacement depth of 6 miles, the resulting reduction in pressure produced numerous sets of vertical fractures, making the granite bodies highly susceptible to the process of rockfall. Several million years later, world temperatures dropped and mountain glaciers entered Yosemite Valley, producing all the elements necessary for today's scenery.

John Muir, the celebrated nineteenth-century naturalist, originated the concept that Yosemite Valley had evolved exclusively through glacial erosion. Detractors countered that a catastrophic earthquake or the simple action of flowing water was the culprit of change. Still others argued that mass wasting associated with the numerous fractures had the greatest impact. Today, combinations of flowing water, glacial erosion, and ice-induced rockfalls are considered the principal forces that have sculpted the landscape. The brute mass of El Capitan, however, remains distinct, because its largely unfractured bulk is relatively immune to the destructive forces of rockfall.

El Capitan, the recognized standard for big-wall ascent, continues to be a mecca for wannabe climbers intent on testing their mettle. Those disinclined to cliff-clinging thrills can scale the crest of El Cap using the trail next to Yosemite Falls. At the top hiker and climber can stand side by side and gaze with awe across one of the most celebrated and picturesque valleys on Earth. Controversy aside, few who have experienced this sight have ever taken it "for granite."

The knife-cut face of the heavily fractured Half Dome, shaped largely by rockfall, offers geometric contrast to the relatively fracture-free bulk of El Capitan.

Artistic interpretation showing the "map of North America" intrusion (right) as well as a linear dike that cuts across the massive face of El Capitan.
—Drawing by Patsy Faires, Kernersville, NC

The "map of North America," to the right of El Cap's Nose, adds a sense of geography to this 3-billion-cubic-foot rock brute. —Courtesy of Erik Skindrud, National Park Service

12. Boulder Flatirons, Colorado

39° 59′ 56″ North, 105° 16′ 58″ West
Phanerozoic Eon Mountain Building

The history of the Rocky Mountains is preserved in the rocks underlying these five photogenic giants.

"Turn! Turn! Turn!" the often-quoted song presented by the rock group The Byrds in 1965, based on verses from the book of Ecclesiastes, includes the phrase "a time to build up, a time to break down." In Colorado, another "rock narration," centered upon the five ramparts that identify the Front Range of the Rocky Mountains outside Boulder, offers dramatic evidence of just such a time occurring during a long span of geologic history.

A time to build up: For 3,000 million years Earth experienced its rites of passage of birth, adolescence, and middle age. A brittle crust enveloped a near-molten mantle, surrounding a molten core. Volcanoes belched forth lava and ash, and earthquakes transformed the early topography. Once the early poisonous atmosphere was neutralized, life responded positively. Critters crawled from the oceans, scampered about the protocontinents, and assumed amphibian habits through evolution. Proto–North America slowly collided with proto–South America and Africa, resulting in a tectonic fender bender—the uplifting of the ancestral Rockies—that stretched from Wyoming to New Mexico.

A time to break down: Rivers transported floods of debris off the ancestral Rockies. The first deposit, the Fountain Formation, lay directly on the Precambrian basement. The deposition of three more formations followed: the Lyons, representing desert conditions marked by sand dunes; and the Lykins and Morrison, both floodplain deposits. As the Morrison Formation was deposited, both herbivores and carnivores sallied across the marshy lowlands in search of mates and food. Preserved footprints and fossilized bones of more than seventy species attest to this being the Age of Dinosaurs.

Change was under way. Throughout the Cretaceous period, waters of the Western Interior Seaway advanced and retreated, altering the environment from terrestrial to marine and resulting in three more sequences of strata, together measuring 10,000 feet thick. The Benton and Pierre formations, mostly shale deposited in offshore conditions and sands deposited in nearshore waters, are separated by the Niobrara, a limestone deposited in deep water. Around 65 million years ago generous volumes of magma invaded and uplifted the Precambrian basement, along with the overlying, younger sedimentary rock, which was bent into complex folds that extended for hundreds of miles. Gold, silver, lead, and tungsten precipitated from the hot waters of the magma and formed the world-class Colorado Mineral Belt. Dinosaurs died out and mammals dominated, setting the stage for the evolution of humankind. The modern Rockies were born.

Over time, erosion partially removed the blanket of younger strata from the cap of the Rockies, leaving the Precambrian granite core flanked by the remains of the sedimentary cover. A juxtaposition of topographic high and low identified the Front Range, the geographic contact of prairie and mountain. Finally, river torrents created adjoining canyons in Boulder (and other areas along the Front Range), sculpting the massive layer of red-hued Fountain Formation into triangle-shaped ramparts. The 50-degree angle of the resulting five Flatirons is ample and overwhelming evidence of the prodigious power of continental-scale mountain building, considering that the Fountain Formation had been deposited in horizontal layers. The soaring, imposing, and stark slablike appearance of the Flatirons is nature's way of emphasizing this dramatic story of long-term geologic development. The Boulder Flatirons are the true rock-stars of time and geologic processes.

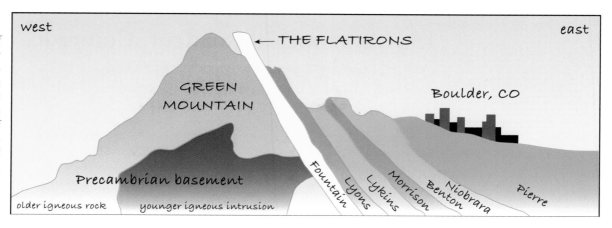

A sheath of sedimentary rock formations flanks the igneous Precambrian basement of Green Mountain. (Modified after Runnells, 1980.)

The Western Interior Seaway, one of the greatest epicontinental seas of all time, extended through the entirety of North America.

The Boulder Flatirons form the east flank of Green Mountain.
—Courtesy of John Karachewski, www.geoscapesphotography.com

43

13. Interstate 70 Roadcut, Colorado

39° 42' 13" North, 105° 12' 05" West
Mesozoic Era Mountain Building

These rocks tell the tale of continental-scale environmental change in a dinosaur-dominated world.

The construction of the 46,726-mile-long Interstate Highway System, which began in 1956, was a boon to the geologic community. Before that, cross-country routes either avoided mountainous terrain or crossed it with costly switchbacks. In contrast, the interstates conquered hilly territory by frontal attack and in the process created roadcuts, cross sections of naked rock flanking many roads. Today, hundreds exist throughout the United States. One of the best known is the I-70 roadcut at exit 259, immediately west of Denver, where the prairie of the Midwest meets the majesty of the modern Rocky Mountains. This national point of geologic interest is designed with convenient parking lots and adjacent walkways protected by concrete posts that act as reference points. There are seventy-eight posts along the walkway on the north side of the interstate.

In unabashed exposure, these sedimentary rocks tell a tale of 140 million years of history. Within this story of change—the only constant in geologic history—is a stage focused on a dinosaur-dominated world set against a backdrop of an inland sea of continental proportion. The interpretation begins by recognizing that these rocks are tilted: they dip to the east and beneath the city of Denver. According to the principle of original horizontality, these layers initially lay in a flat position. Their angle is the result of mountain building forces that occurred millions of years after they were deposited.

North-side posts 1 through 38 identify the 161-to-145-million-year-old Morrison Formation, composed of gray, green, and red sandstone and shale. Deposited under environmental conditions ranging from meandering rivers and floodplains to scattered marshland, Morrison strata stretch across a multistate area and gained international fame in 1877 with the discovery of dinosaur remains. Though there are no fossils in evidence at this geo-site, the bones of more than seventy species have earned these rocks the designation the "graveyard of the dinosaurs." Probably the most famous is the 40-ton, long-necked apatosaurus (formerly brontosaurus), the definitive faunal linebacker of its time.

Posts 38 through 78 identify the 145-to-99-million-year-old Dakota Group, composed of gray, yellow, and pinkish sandstone and shale. This group is a composite of two formations: the Lytle (posts 38–51), which was deposited by rivers; and the South Platte (posts 51–78), reflecting, especially in the dark gray rocks at post 56, the change to a marine environment. The South Platte rocks represent the encroachment of the Western Interior Seaway, a 600-mile-wide ocean that extended east to Minnesota and from the Arctic Ocean to the Gulf of Mexico. (The symmetrical ripple marks exposed 1 mile south, along Hog Back Road, or Colorado 26, along with dinosaur footprints existing elsewhere in the formation, provide evidence of intertidal wave action across a beach environment.) Finally, the youngest rock units of this exposure, sandstones associated with the upper portion of the Dakota Group (posts 65–78), are the units that contain large nearby reserves of oil, helping make Colorado the number eleven producer of oil in the United States.

The Western Interior Seaway dominated this scene until 70 million years ago, when earth-scale compressional forces began to form the modern Rocky Mountains, and concurrently uplifted the I-70 roadcut strata to an angle of 45 degrees. This former seascape and mountain building event created for today's interstate traveler a geologic story of exceptional clarity.

The color change near post 56 between sandstone units (left) and shale units of the South Platte Formation marks the beginning of the invasion of the Western Interior Seaway, in which the finer-grained sediment of the shale was deposited.

A panorama of color and angle highlights the I-70 roadcut.
—Courtesy of Vincent Matthews, Colorado Geological Survey

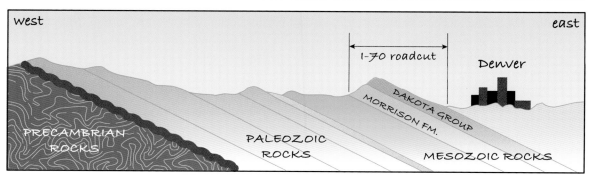

To the west of Denver, Morrison Formation and Dakota Group rocks overlie older Paleozoic-age and Precambrian-age rocks.

14. Florissant Fossil Beds, Colorado

38° 54' 48" North, 105° 17' 08" West
Paleogene Period Fossilization

Fossil tsetse flies and stinkbugs reveal how a 35-million-year-old world differed from the present.

Florissant Fossil Beds National Monument, immediately south of Florissant, Colorado, off US 24, offers a most unusual and multifaceted insight into a former life. Here, paleontologists have collected 1,700 species of insect and plant life from the brown shale and volcanic ash layers of the middle Paleogene-age Florissant Formation. More than 130 years of research have revealed evidence of a climate as much as 25 degrees Fahrenheit warmer and a rainy season 30 percent wetter than today.

The 65-mile drive west along US 24 between Colorado Springs and Hartsel, Colorado, rises 9,507 feet to Wilkerson Pass and extends across some of the most pastoral alpine scenery in the central Rocky Mountains. Several tens of millions of years ago, the same route, then a terrain of youthful volcanic rock that had welded to a basement of granite, was fraught with lethal danger, not to any humanoids, for they had yet to evolve, but to the myriad swarms of insects and plants that called the region home. Volcanic activity was alive and shouting out loud. Where Mount Princeton presently towers over central Colorado, the Guffey volcanic complex was then erupting and producing prodigious ashfall, giving birth to one of the most outstanding fossil depositories in the world, rivaled in diversity and richness only by the Solnhofen limestone quarry of Germany and the Baltic amber district of northern Europe.

Thirty-six million years ago, after the Guffey complex had awakened from one of its periodic slumbers, a mudflow traveling down the slopes of one or more of the complex's volcanoes inundated a forest of giant sequoia trees. Several of these petrified stumps have been exhumed by erosion, but more than eighty remain buried in the extensive meadow fronting the visitors center at the national monument. A subsequent mudflow formed a natural dam across one of the region's ancient valleys and in the process created a lake 1 mile wide by 12 miles long. It existed more than 2 million years, frozen by alpine temperatures in the winter and then covered by diatomaceous blooms in the heat of summer. These mucous blooms were sticky enough to be deleterious to any insect or leaf settling on their surface. Combined with fallout from ash clouds, the blooms created a muddy slurry that settled to the lake floor and then hardened to paper-thin shale strata that average only 0.02 inches in thickness. These strata became the Florissant Fossil Beds.

Summer after summer the cycle of sedimentation entrapped volumes of leaves, seeds, flowers, pollen, and cattails in intimate association with butterflies, bumblebees, wasps, beetles, stinkbugs, tsetse flies, cicadas, cockroaches, termites, ants, lacewings, mosquitoes, and midges. The state of preservation of some of these fossils is most unusual. In one fossil the compound eye of a fly is clearly delineated; in another an assassin bug is frozen in time in the act of feeding on a leaf.

The cyclical biologic change and geologic activity centered around this ancient lake, now the Florissant valley, created a distinctive window into the recent past—a true Rosetta stone to the Paleogene world. Today, because of erosion over the last 30 or so million years, only vestiges of the former lake remain, in the form of the fossiliferous Florissant Fossil Beds. While the 1-mile-long Petrified Forest Loop that crisscrosses the monument's landscape gives up close and intimate access to the fossil beds, as well as a view of an excavation site, fossils are best seen in the exhibits in the visitors center. The biotic diversity and beauty of these lake remnants could not have been more appropriately named, since *florissant* is French for "flowering" or "blooming."

Big Stump, with its 38-foot circumference, is evidence of the many Sequoia trees that once forested central Colorado.

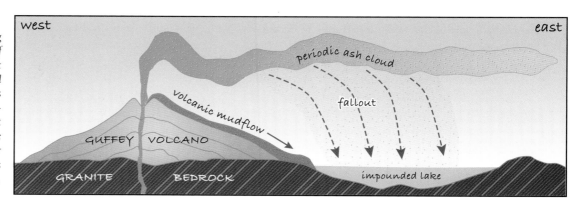

Mudflows flowing down volcanoes of the Guffey volcanic complex entombed trees up to heights of 15 feet and created a lake that became an effective depository for numerous insects and plants.

A wasp (right) and a lacewing insect, both approximately 1 inch long, are some of the insects preserved in lifelike detail in Florissant shale and ash strata. —Drawings by Patsy Faires, Kernersville, NC

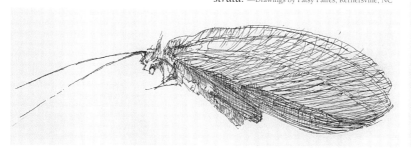

15. Dinosaur Trackway, Connecticut

41° 39' 07" North, 72° 39' 25" West
Mesozoic Era Fossilization

Follow the footsteps of the largest carnivore of its time as it roamed the land in search of prey.

Beneath the geodesic dome at Dinosaur State Park, in Rocky Hill, 1 mile east of exit 23 off I-91, is a one-of-a-kind exhibit where five hundred toe and claw imprints are preserved in situ within sandstone deposited along an ancient shoreline. As early as 1841, scholars argued that a bird with feathered legs or an extinct race of giant chickens had produced the tracks. Today, having compared them to similar imprints from Texas (see geo-site 80), many scientists believe the tracks were made by *Dilophosaurus*, a two-crested, all jowl and jaw, meat-eating dinosaur that flourished during the transition from the Triassic to the Jurassic period.

Twenty feet long, this carnivore weighed in at 1,000 pounds, walked upright, stood 8 feet tall, and took first dibs at not only lunchtime, but all other times, using its fifty-six-toothed-jaw. The haphazard track pattern at Dinosaur State Park suggests it hunted and functioned as an individual, not as part of a herd. While it was geologically short-lived, *Dilophosaurus* is an important link in the evolutionary chain that produced the great *Tyrannosaurus* and *Brachiosaurus* genuses of dinosaur that dominated Earth during the Cretaceous period.

The page of geologic history that records the change from the Paleozoic era to the Mesozoic era—before *Dilophosaurus* came to be—is earmarked by an event of first magnitude. After almost 300 million years of being jostled about the globe by plate tectonic process, the early landmasses had finally come together as the supercontinent Pangaea. Across the width and breadth of this one Earth, which extended from pole to pole, a physical calm prevailed, and for good reason. All the forces that could alter the topographic nature of land by creating folds, faults, earthquakes, and mountain systems had temporarily run their course.

The situation within the biologic world was not so tranquil. By the end of the Paleozoic era as many as 95 percent of ocean-dwelling invertebrate species and 70 percent of their air-breathing brethren had ceased to exist. The cause, whether volcanic eruption, meteor impact, or habitat consolidation, is still under debate. While this extinction event came very close to clearing the slate, it did set the stage for the reign of dinosaurs, including *Dilophosaurus*.

During this time of transition Connecticut was located 20 degrees north of the equator. Lakes dotted its countryside and forests of ginkgo, cycads, and conifers flourished in an arid and monsoonal environment. Beneath this tranquil setting, gargantuan, upwelling pods of deep-seated magma began to bull their way up through the underside of Pangaea. As the pods of magma neared the surface, they spread laterally, stretching and ultimately fracturing Pangaea into the seven masses that are today's continents. These stretch marks exist today as a linear zone of twenty sandstone- and shale-filled half grabens that extend from Florida north to Nova Scotia. A half graben is a rift in the crust that is bound on one side by a fault. The Hartford Basin of central Connecticut is such a geologic feature.

The Hartford Basin is filled with 5,000 feet of red-bed sediment deposited in floodplains and rivers and three 200- to 450-foot-thick lava flows. With its preserved treasure of fossil tracks, the Hartford Basin is a depository of one very dramatic chapter in the rise-and-fall story of the dinosaurs that trod the evolutionary stage during the Mesozoic era. Dinosaur State Park, lauded as one of the country's largest concentrations of tracks, displays the world of *Dilophosaurus*, one species of "terrible reptile" that after millions of years near the top of the food chain fell prey to that great biologic equalizer—extinction.

At Dinosaur State Park, intersecting sets of pockmarking footprints give evidence of a well-traveled freeway.
—Photograph © by Mark Chesner

Geologic map of Connecticut and a cross section showing the lava beds and sedimentary rock that fill the Hartford Basin, a half graben. Older metamorphic rock forms the geologic basement.

Two hundred million years ago a dinosaur, perhaps the carnivorous Dilophosaurus, *left this three-toed imprint as evidence of its size and weight. Pen for scale.* —Courtesy of Kelly Lowder, Alexandria, VA

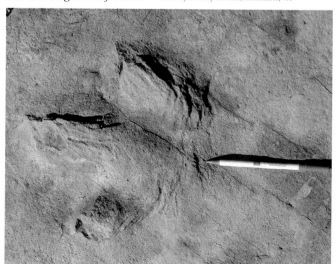

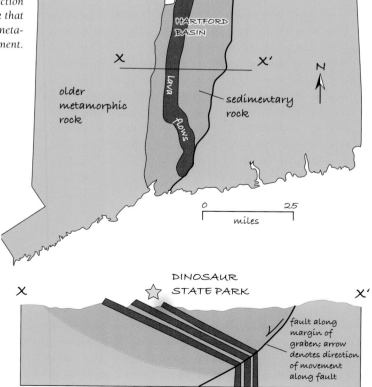

16. Wilmington Blue Rocks, Delaware

39° 49' 00" North, 75° 34' 09" West
Ordovician Period Metamorphism

Colorful rocks that relate both to the birth of the Appalachian Mountains and to major league baseball.

Carved from a metamorphosed terrane bearing evidence of the birth pangs of the Appalachian Mountains, the rolling landscape of northern New Castle County, Delaware, is home to the famed Wilmington "Blue Rock" Complex, also known as the Brandywine Blue Gneiss.

Approximately 500 million years ago, Laurentia, the landmass that would become North America, was 40 percent smaller than the continent is today and turned almost 90 degrees clockwise from its current orientation. The entire continent lay between 15 degrees north and 30 degrees south latitudes. The smaller landmass of Taconica, a 1,500-mile-long arc of volcanic islands, punctuated the waters of the Iapetus Ocean south of Laurentia, and a subduction zone floored the body of water that separated these two landmasses.

As the leading edge of Laurentia slowly descended into the depths of the mantle in the subduction zone, its uppermost surface warmed to the melting point and formed a magma chamber that added to the volcanic rock composing Taconica. The curtain opened on the first chapter in the construction of the Appalachian Mountains, and the stage was set for the ultimate destruction of the Iapetus Ocean some 300 million years later.

Over the ensuing 50 million years the two landmassses converged. First, the island arc scraped off the veneer of marine mud and sand that had accumulated on the seafloor of the descending Laurentia Plate. This sediment did not go deep enough into the subduction zone to be melted, but it was buried deeply beneath Taconica nonetheless, where it was subjected to intense temperature and pressure to form metamorphic rock. During Late Ordovician time these tectonic events reached a climax: the portion of the Iapetus Ocean between the two landmasses closed, the plates collided, and the volcanic arc overrode the pile of oceanic sediment. During this cataclysmic collision, the volcanic rocks and the oceanic sediments were severely metamorphosed.

This welding of landmasses—called the Taconic orogeny—created an intensely folded and faulted Himalayan-scale mountain chain that stretched from present-day Nova Scotia to Georgia and extended the eastern border of proto–North America (the southern coast at the time) some 100 miles. Over the next 200 million years, two other periods of mountain building created the modern Appalachian mountain chain. Today, only the eroded roots of Taconica's former heights remain.

Impressive rock exposures representing this titanic clash of continent and island arc are easy to find in Brandywine Creek State Park, off Thompson Bridge Road 6 miles north of downtown Wilmington. On the hillside between the parking lot and the footbridge over Rocky Run are large outcrops that belong to the Wissahickon Formation, the metamorphosed remains of sand and mud from the Iapetus Ocean that was squeezed shut. These rocks are distinguished by their light color and the presence of black mica and garnet crystals—some up to 0.75 inch in diameter—that have a blocky nature. Approximately 500 feet south of the bridge, darker masses of the Wilmington "Blue Rock" Complex make their appearance. These metamorphosed volcanic rocks weather to rounded form and feature quartz-rich, light- and dark-colored bands up to several feet thick—the hallmarks of the metamorphic rock gneiss. This zone of gargantuan tectonic impact celebrates the head-on collision of continental-sized plates during a time when Earth geography was undergoing major reconstruction.

When fractured, the quartz of the Wilmington Complex often displays inclusions (fragments of older rock within an igneous rock to which they may or may not be genetically related) that vary from bluish gray to bright royal blue. Named "blue rocks" by nineteenth-century quarrymen, they are so closely associated with the city that its baseball team—a Class A affiliate of the Kansas City Royals—is affectionately named the Wilmington Blue Rocks.

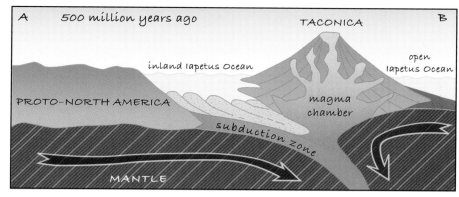

The convergence of proto–North America (Laurentia) with Taconica slowly closed the inland portion of the Iapetus Ocean and began to alter both volcanic rock (green) and ocean sediments (yellow) to their metamorphic equivalents. Arrows denote direction of plate movement.

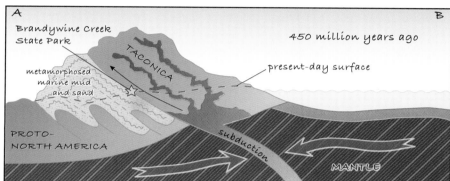

The Taconic orogeny climaxed when the collision of Taconica and proto–North America completed the metamorphism of ocean sediments into the Wissahickon Formation (yellow) and volcanic rocks into the Wilmington "Blue Rock" Complex (green). Arrows denote direction of plate movement.

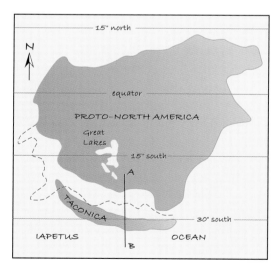

The geographic setting of proto–North America (Laurentia) and Taconica 500 million years ago. The present-day east coast of the United States (dashed line) and the Great Lakes give perspective. The line labeled "A" and "B" represents the land portrayed in the cross sections above.

Sunlight on a freshly broken exposure of the Wilmington "Blue Rock" Complex enhances its acclaimed bluish gray color. —Courtesy of the Delaware Geological Survey

17. Devil's Millhopper, Florida

29° 42' 48" North, 82° 23' 31" West
Cenozoic Era Karst Formation

In form and formation, the model for the hundreds of sinkholes that riddle the bedrock of Florida.

The outer 10 miles of the crust of Earth consists of 95 percent igneous and metamorphic classes of rock, with the remaining 5 percent sedimentary. When it comes to the very surface, however, the percentages are almost the reverse: approximately 75 percent of all exposed bedrock on the seven continents is sedimentary. In short, while quite small in relative volume, sedimentary rock dominates the continental surface of the Earth.

Sedimentary rocks can be divided into two classes. The detritus category—those formed by the consolidation of solid particles—includes conglomerate, sandstone, and shale. Chemical types, commonly catalogued as limestone, form in two ways: by inorganic chemical means, such as the precipitation of minerals out of seawater, and when the remains of minute aquatic organisms are deposited on the seafloor, eventually lithifying as solid rock. The latter biogenic category accounts for 90 percent of all limestone.

Fifty million years ago, the present-day region of the state of Florida was home to myriad minute organisms thriving in shallow, sunlit marine waters. When they died, their calcium carbonate skeletal remains gradually settled on the seafloor in great thicknesses, forming limestone that became the geologic foundation of the Sunshine State.

Riddled with fissures, crevices, and cracks, these strata function as a major recharge zone for the Florida Aquifer, one of the world's most prolific reservoirs of groundwater. In a world increasingly deficient in potable water, that is the good news. However, at shallow depths these same rocks are prone to karst-type weathering and erosion, which creates a topography characterized by sinkholes, caves, and underground drainage. That is the bad news, because this style of landscape causes damage to property, utilities, and roads, not to mention being associated with groundwater pollution.

While limestone can be brittle and hard, it also can be easily dissolved by naturally occurring acidic solutions. When precipitation, in the form of fog, mist, and rain, is exposed to an atmosphere laden with carbon dioxide, it is altered to a weak carbonic acid. This solution can become even stronger after it soaks into the ground and comes into contact with soils that contain decaying organic material. In the presence of this potent acid, limestone slowly dissolves, initially creating a vug (a small hollow), then a cavity, and finally a cavern. As the cavern increases in size, its roof decreases in thickness until it collapses, forming a sinkhole.

A classic example of this type of erosion, easily explored by way of a stairway that descends to its cave-ridden bottom, is seen at the Devil's Millhopper Geological State Park, in the northwest suburbs of Gainesville, Florida. This beautifully preserved, circular and steep-sided sinkhole measures 120 feet deep by 500 feet in diameter. It formed when a cavern dissolved in 37-million-year-old Ocala Limestone collapsed, but its exact age is problematic. Based on the presence of fossil plant and animal remains, it is estimated to be between 10,000 and 15,000 years old. The distinctive name derives from its unusual funnel-like shape—reminiscent of the early devices used by local farmers to grind their grain—and the presence of preserved bones and teeth supposedly related to past devilish activities.

All of Florida is susceptible to the development of karst, especially those regions where limestone lies close to the surface (orange).

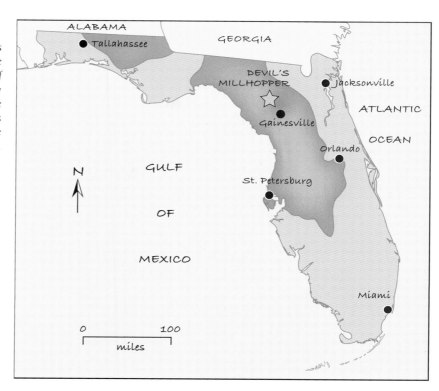

These 1- to 2-foot-wide pits dissolved by carbonic acid typify many karst terrains.
—Courtesy of Tom Scott, Florida Geological Survey

The collapse of the Winter Park Sinkhole swallowed a three-bedroom home, five Porsches, and half an Olympic-sized pool. —Courtesy of Frank Rupert, Florida Geological Survey

18. Stone Mountain, Georgia

33° 48' 20" North, 84° 08' 44" West
Pennsylvanian Period Exfoliation

Whether it's called a monadnock or an exfoliation dome, this bald-headed massif is famous for its smoothed profile.

The distribution of land and water across the surface of Earth cannot be understood until that most basic of questions is addressed: why are the continents dry and the ocean floors wet? The answer is simple: the continents are rooted in granite, a light-colored rock with a density of 2.7 grams per cubic centimeter (g/cm^3), and the ocean floor is composed of basalt, a dark rock with a higher density of 3 g/cm^3. Basalt sits low in the crust of Earth and acts as a collection basin for water as it follows its natural instinct to find the lowest level. In contrast, the lighter granite stands high—and is thus dry.

Throughout geologic history, much like the insides of massive lava lamps, columns of molten granite have been thrust upward through the bowels of Earth's crust, crystallizing at great depths. The resultant plutons are coarse grained, pink to gray in color, and rich in two defining minerals: quartz and feldspar.

Outside Atlanta, 14 miles east of the intersection of I-75 and I-85, the 1,686-foot-high promontory of Stone Mountain, billed as the world's largest exposed piece of granite, dominates the countryside. A textbook example of a monadnock, a hill that stands above the surrounding countryside, Stone Mountain measures 1.8 by 0.8 miles and stands 780 feet above the metamorphic terrain of the enveloping Piedmont Province. Radiometric age dating indicates the granite is approximately 300 million years old. Its globular shape is due entirely to exfoliation, a weathering process wrapped in a history of controversy.

For decades, geologists thought temperature change caused exfoliation, in essence, the separation of concentric plates of rock from a large rock mass. They assumed expansion and contraction of a rock surface due to daily heat fluctuations weakened the mineral fabric, and that fractures developed to relieve the resulting differential stress. Modern science suggests temperature plays only a minor role in exfoliation.

Today, it is known that Stone Mountain evolved principally through the exfoliation that occurs in rock that was once deeply buried but has since been brought to the surface through erosion, thus releasing its confining pressure. In the final mountain building stage of the Appalachian Mountains, granitic magma reached depths of 40,000 to 50,000 feet and then crystallized. Through the ensuing millions of years the overlying multimile-thick column of country rock eroded away, exposing the plutons. Without the confining pressure of overlying rock, granite plutons such as Stone Mountain expanded in volume while their shapes stayed the same. This increase in mass created fractures oriented parallel to the surfaces of the plutons, separating sheets of rock much like an onion peels apart. Eventually these sheets separate completely and fall away, leaving the rock body smoothed and rounded. This type of weathering, whereby an exfoliation dome forms, is found where igneous rock abounds, most particularly where granite is common.

In addition to separated sheets of granite, other features endemic to exfoliation domes are present at the east-side Quarry Exhibit at Stone Mountain: xenoliths, fragments of the surrounding country rock incorporated into the magma body as it intruded the region; flow banding, magma flow patterns highlighted by the presence of platy, translucent flakes of muscovite; tourmaline pods (also known as "cat's paws"), silver dollar–sized clusters of black tourmaline crystals surrounded by a white halo; and granite dikes, injections of magma that passed through the pluton during late stages of crystallization.

Stone Mountain exists today because its constitution of dense granite—commercially the stuff of tombstones and kitchen countertops—is more resistant to erosion than the rocks it intruded. Optimal for sculpting, Stone Mountain is the site of a well-known Confederate memorial.

The colossal igneous mass of Stone Mountain measures 7 miles in circumference and exposes 20 billion cubic feet of granite. —Courtesy of Stone Mountain Memorial Association

Weathering processes continue to strip away rock layer after rock layer of Stone Mountain, rounding its profile and softening its edges. Pen for scale.

Raisin-in-the-pudding xenoliths are pieces of preexisting rock that were incorporated into the Stone Mountain magma as it rose toward the surface. This dark chunk of biotite gneiss is metamorphic rock. Pen for scale.

19. Kilauea Volcano, Hawaii

19° 23' 21" North, 155° 06' 22" West
Neogene Period Intraplate Volcanism

Active since 1983, this crater is a wide-open window into the hellish depths of a full-blown hot spot.

According to plate tectonics theory, some 95 percent of the world's active volcanoes lie along the edges of the fourteen or so large and small plates that comprise Earth's lithosphere. The remaining 5 percent commonly form as elongate chains great distances from plate boundaries. The Hawaiian Islands, 2,000 miles from the nearest plate border, are illustrative of this intraplate category of volcanism.

Cited as the most energetic volcano on Earth, Kilauea has ejected on average several hundred thousand cubic yards of lava daily since 1983 and carries title to the longest eruption in recorded history. It is the youngest of the five shield volcanoes—flattened domes built of very fluid lava—that have fused together as the Big Island of Hawaii. Even more significantly, the Big Island is the youngest of the Hawaiian Island–Emperor Seamount chain of more than eighty islands and numerous seamounts—all former volcanoes—that extends more than 3,700 miles across the Pacific Ocean, a distance greater than the width of the continental United States. The oldest seamount is at the opposite end of the chain, close to the Aleutian Trench, an oceanic depression at the subduction zone off Alaska. This geographic separation of youngest volcano from oldest seamount is key to understanding how long chains of intraplate volcanoes develop.

In the early 1960s, when plate tectonics theory was being subjected to intense and universal scrutiny, new evidence suggested that intraplate volcanoes were related to localized sources of heat, called hot spots, that originated deep within the mantle. In the case of the Hawaiian Island–Emperor Seamount chain, the heat from such a hot spot partially melted the overlying Pacific Plate. Driven by buoyancy, this magma slowly rose through the crust and eventually erupted onto the Pacific seafloor in the form of a seamount (a submerged volcano). Countless eruptions later, having grown in mass and height, the seamount emerged above the waves as an active island volcano.

Simultaneously, the Pacific Plate was moving to the northwest, carrying this first-formed island beyond the stationary hot spot. Without a life-giving supply of magma, the island aged, first to dormancy then extinction. In time, a newer, younger volcano developed over the hot spot, the second link in a growing chain. This sequence of birth and death was repeated, again and again and again. Over the course of the past 70 million years an extended trail of islands and seamounts has left its mark across the Pacific Ocean. Kilauea, the latest volcanic island to form, began life as a seamount some 500,000 years ago, and 400,000 years later emerged a full-fledged island. The next link of the Hawaiian Island–Emperor Seamount chain history is being written 20 miles off the southeastern coast of Hawaii. Active since the mid-1990s, the 10,000-foot-high seamount Lo'ihi is dependent on the very same 40-mile-deep plumbing system that continues to nourish its big sister Kilauea.

In the East Rift sector of Kilauea, streams of molten rock periodically erupt from the Pu'u O'o vent and flow through a 6-mile-long tube system into the sea. There, the old real estate sales pitch to "buy land because it isn't being made anymore" is proven erroneous, in a shroud of steam. Home to Pele, the goddess of fire, Kilauea continues to maintain its reputation as one of the few areas on Earth where the fiery forces of volcanism can be inspected on an up close and very personal basis.

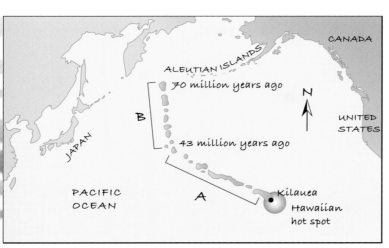

The Hawaiian Island (A) and Emperor Seamount (B) chain extends more than 3,700 miles from its point of origin to the current location of the hot spot beneath Kilauea. The kink in the chain formed 43 million years ago when the Pacific Plate changed direction.

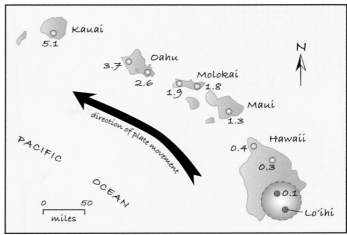

The hot spot today nurses the Lo'ihi seamount as it did the older volcanoes in the past. Numbers represent age of volcanic rocks in millions of years.

Five volcanoes form the island of Hawaii, as shown in the small map. Two rift zones intersect the youngest volcano, Kiluaea (number 1). Present-day eruptions on Kiluaea are centered at the Pu'u O'o vent (inset photo). —Inset photo courtesy of Jane Takahashi, U.S. Geological Survey Hawaiian Volcano Observatory

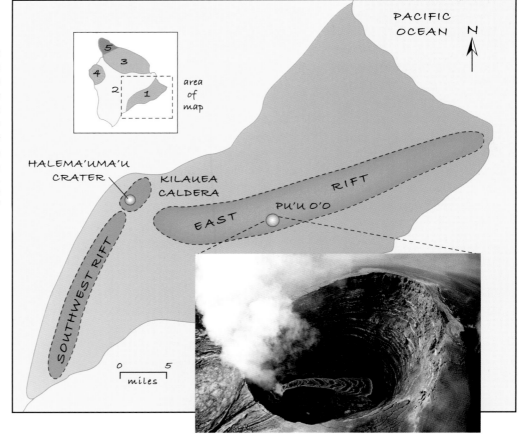

20. Borah Peak, Idaho

44° 09' 50" North, 113° 52' 04" West
Holocene Epoch Earthquake Effect

This fault rupture pushed a mountain range up almost 12 inches and caused a valley to collapse 8 feet.

On Friday morning, October 28, 1983, tectonic disturbance shattered the tranquility of the Lost River Valley for the first time in 15,000 years. At precisely 8:06, shock waves radiated outward in a three-dimensional pattern at a velocity approximating 4 miles per second. The event lasted ten seconds short of a full minute—a lifetime to the residents of Arco, Mackay, and Challis. Within minutes, eight western states and two Canadian provinces—from Salt Lake City in the south to Calgary in the north—felt the shock waves. Within the hour, the scientific community had reported the heart of this catastrophic burst of energy lay 7 miles below a nondescript plot of rangeland 19 miles northwest of the old mining town of Mackay.

With a magnitude of 7.3, the energy released by the Borah Peak earthquake (estimated to equal the explosive force of 1.5 million tons of TNT) registers between that of the 1945 Hiroshima atomic bomb explosion (13,000 tons of TNT) and the 1980 eruption of Mount St. Helens (24 million tons of TNT). The effects of the tremor were instantaneous. The rock-solid limestone surface above the epicenter snapped, giving birth to a 21-mile-long scarp that ripped along the southwestern shoulder of the Lost River Range, sounding like a flight of jets at low altitude. Cracks 6 to 12 feet deep rippled through US 93. The 400 billion gallons of groundwater that gushed from newly opened ground fissures doubled the normal discharge of Big Lost River. One 50-ton boulder careened down from its mountain height, and 150 miles to the east Old Faithful Geyser in Yellowstone National Park temporarily became unfaithful.

The Borah Peak fault scarp is starkly exposed along Doublespring Pass Road, 2.5 miles northeast of US 93 turnoff at mile marker 131.5. Here, adjacent to a parking area, the fault zone consists of a series of individual, stair-step displacements, totaling approximately 115 feet in width. Within easy walking distance, the principal fault scarp varies from 6 to almost 16 feet in height—eureka-moment evidence that in nature the constructive forces of mountain building are in everlasting combat with the destructive forces of erosion.

When rock is subjected to applied force, it undergoes strain, a change in either shape or volume. Excessive force, whether it's compression or tension, typically causes rock to rupture. When the rock masses on either side of a rupture instantaneously move, the fracture becomes a fault, and the resulting vibration becomes an earthquake. In 1983, Borah Peak, long lauded as the mountain kingpin of Idaho, as well as the crown jewel of the Lost River Range, increased its elevation by almost 1 foot, while the adjacent Thousand Springs Valley collapsed nearly 8 feet. This kind of topographic displacement represents a normal type of fault—one caused by the crust of the Earth being subjected to extensional force.

In this region extensional force is related to the development of the Basin and Range Province, initiated 17 million years ago when the easterly migrating Pacific Plate began to plunge down and under the westerly moving North American Plate at an average rate of 0.5 inch per year. This province, encompassing a large portion of the Intermountain West, is a part of the crust that is rifting apart. The Borah Peak quake is evidence that this pull-apart tectonic activity is perhaps only beginning to make its mark on the topography of southern Idaho. It is but one of the millions of earthquakes that occur each year—evidence that Earth is a pulsating, ever-changing, dynamic, and living planet.

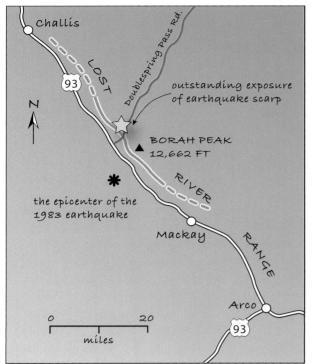

Northwest of Mackay the Borah Peak earthquake scarp is beautifully exposed east of Highway 93. The solid red line indicates the zone of best exposures.

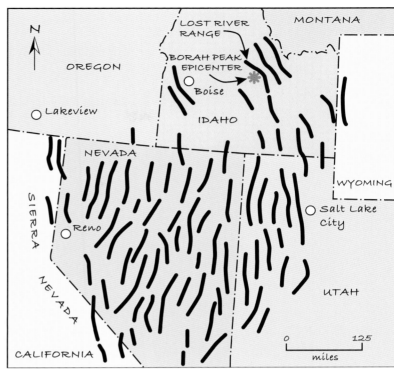

Southern Idaho is part of the Basin and Range Province, in which long, linear mountain ranges rise relative to the dropping valleys between them (black lines).

The stair-step arrangement of the blocks composing this fault zone is evidence of a normal fault, created by forces that fracture and extend the crust.

59

21. Menan Buttes, Idaho

43° 47' 07" North, 111° 58' 31" West
Holocene Epoch Volcanic Activity

Near-twin cones were born under conditions known nowhere else in the conterminous United States.

Volcanism can be defined simply as the surface expression of Earth's internal plumbing system, or more technically as that process associated with the extrusion of gas and molten rock. Active volcanism is widespread: over the ocean floor and on every continent except Australia. While often associated with negative events—as many as sixteen thousand individuals may have been killed in AD 79 when Mount Vesuvius erupted in Italy—this geologic process also lays claim to positive results.

Volcanic gas, principally in the form of water vapor and nitrogen, contributed to the early formation of Earth's hydrosphere and atmosphere. Volcanism increases habitable land—Japan is 100 percent volcanic in origin—and is a renewable source of energy. Iceland, the land of fire and ice, extensively uses geothermal energy derived from volcanism to keep its citizens warm and its industrial engines turning. Volcanism is evidence that Earth is alive.

The most recognizable trademark of volcanism is the cone, a topographic feature that relates to the crust of Earth as acne relates to the skin of a teenager: both indicate youth and vitality. Cones exist in two basic designs—those built of very fluid lava flows and those created by gas-charged eruptions.

Mauna Loa, Earth's largest volcano and one of five that form the Big Island of Hawaii, is a world-class example of a shield volcano, a broad, low-profile, gently sloping giant constructed of low-viscosity lava flows. In contrast, cinder cones, which are small in both size and height and usually steep-sided, are the product of a champagne-popping eruption created by the discharge of viscous, gas-charged lava. Cinder cones are composed of explosively ejected, unconsolidated chunks and blobs of solidified molten rock collectively termed pyroclastic material.

Ten miles southwest of Rexburg, the almost-symmetrical profiles of the Menan Buttes break the flat expanse of sagebrush of southeastern Idaho. Formed through an explosive process involving both magma and water, these phreatomagmatic cinder cones represent the only volcanic eruptions in the contiguous United States that were born through a river. The 60-minute hike up the west flank of the 750-foot-high, publicly owned North Menan Butte is exhilarating and rewarding in terms of the view and the landscape. At the crest, a circular pattern of inward-dipping beds of lava visually identify the 3,000-foot-diameter rim, while in sharp contrast and just a few yards down and along the left side of the well-worn trail, layered beds plunge away in all directions from the central crater.

Ten thousand years ago the bedrock of the Snake River Plain was broken by a fracture that connected to a deep-seated magma chamber that was slowly invading the Earth's crust. Before the vertical column of molten rock could pour onto the surface, however, it developed two prongs that came into contact with a water-saturated, abandoned channel of the Snake River, and the inevitable happened. The instant chilling explosively turned the 2,000-degree-Fahrenheit magma in the twin columns into minute fragments (ash sized) of tachylyte, a type of volcanic glass. As these shards fell back to Earth, forming twin cinder cones, they fused together as a rock called tuff. The resulting Menan Buttes—a unique, textbook pair of "glass" volcanoes—are the largest of their kind, which is found in only a few places in the world. Diamond Head on the Hawaiian Island of Oahu, at a distance of 3,200 miles, is the closest landform with a similar history of development.

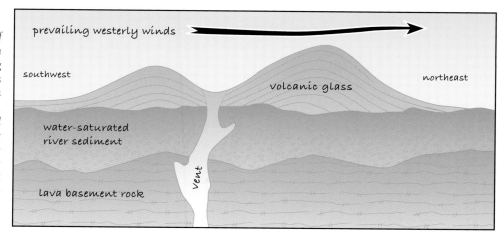

Cross section of North Menan Butte, showing the relationships between the layers of volcanic glass, river sediment, and basement rock. The asymmetry of the cone reflects the prevailing wind direction at the time of the eruption.

An aerial perspective looking down the throat of the 5,800-foot-diameter (at its base) North Menan Butte.
—Courtesy of the Bureau of Land Management

Around the rim of North Menan Butte, dark-toned lava beds plunge inward (to the right) toward the central vent.

22. Great Rift, Idaho

43° 27′ 46″ North, 113° 33′ 38″ West
Holocene Epoch Volcanic Activity

Lunarlike scablands come alive in a crazy-quilt display of rock texture in this bizarre volcanic terrain.

While Hawaii lays claim to the most recent volcanic activity within the fifty states, the Snake River Plain of Idaho is not far behind. Part of this statewide swath of once-molten rock is designated as the Craters of the Moon National Monument and Preserve, home to lava fields as young as 2,100 years in age. Located between Arco and Carey, this area was early on dubbed the strangest 75 square miles in North America, though now the monument and preserve encompasses a much larger area.

If interwoven fields of lava are considered the geologic body of Craters of the Moon, then the Great Rift is its backbone, a more than 50-mile-long, 1-to-5-mile wide system of generally northwest-trending, en-echelon fissures that cross the Snake River Plain. It may actually extend downward to the interface between the crust and mantle of the Earth—a depth as great as 40 miles. The Great Rift is ideally viewed on aerial photographs, but separate elements in the form of aligned arrays of volcanic cones and ground crevasses can be seen from the surface. One crevasse is open to a depth of 800 feet—a rift chasm deeper than any other known on Earth.

The overall extent of the Great Rift is composed of four areas, each containing a set of differently oriented fractures. From north to south, these areas are the Craters of the Moon set (oriented northwest at 35 degrees), the Open Crack set (northwest at 30 degrees), the Kings Bowl set (northwest at 10 degrees), and the Wapi set (oriented north-south). The Wapi fractures are partially covered by a massive lava flow of the same name, while the Kings Bowl and Open Crack sets are characterized by yawning and precipitous fractures, the greatest example being Crystal Ice Cave, which is perennially shrouded in ice. The Craters unit, the largest of the four, is largely covered by approximately sixty individual lava flows—one 45 miles long—that envelop an area of 600 square miles.

Throughout its extent, the Great Rift functions as a conduit for the extrusion of molten rock from deep underground. While most of this activity is nonviolent, the course of the Great Rift is marked by diverse features, such as caves and cinder cones, which have been born by molten rock vomiting, burping, spitting, and sneezing its way upward from the depths.

Eight cycles of volcanic activity have been mapped in the Craters of the Moon unit. The opening act of each cycle typically involves a curtain of fire display of very fluid lava that can soar to several hundred feet as it erupts from fractures. With the passage of time, the pyrotechnic curtain breaks into isolated vents of erupting lava. A low, steep-sided spatter cone composed of pasty blobs of lava may form around any vent that emits gas-rich lava. A vent that ejects lightweight, vesicular fragments generally creates a cinder cone (see also geo-site 21). During the concluding act, volumes of very fluid lava spew out of the vents and flow across the countryside.

Other aspects of rifting are associated with the Snake River Plain, but the Great Rift has earned the blue ribbon as the largest, deepest, and youngest volcanic rift system in the conterminous United States. Its presence is evidence that the basement structure underlying southern Idaho is continually interacting with tensional pull-apart tectonic forces that continue to rip this portion of the Earth's crust asunder. Detailed studies show that each eruptive cycle lasts on average 2,000 years. Since the last cycle ended 2,100 years ago, the next rip in the crust is an event just waiting to happen. At that time, the Great Rift will become even greater.

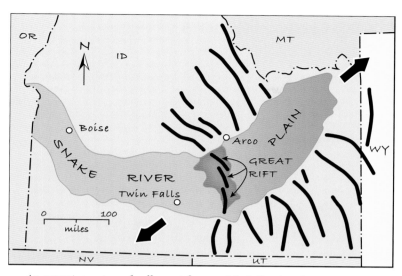

An extensive system of pull-apart fissures (black lines) underlies the Snake River Plain (green). One segment—the Great Rift—pierces the surface of the Craters of the Moon National Monument (orange).

Steam explosions created the 300-by-100-foot Kings Bowl when ascending magma came into contact with groundwater.
—Courtesy of the National Park Service

The alignment of cinder cones in the foreground with the one on the horizon gives geographic definition to the overall trend of the Great Rift.
—Courtesy of the National Park Service

23. Valmeyer Anticline, Illinois

38° 17′ 48″ North, 90° 18′ 28″ West
Paleozoic Era Basin Formation

One small anticline offers clues to the development of the 350-million-year-old Illinois Basin.

The drive from Indianapolis to St. Louis is for many motorists an experience of sameness blending into more sameness. The hard-rock geologist, however, sees this undulating landscape through different eyes, knowing that beneath the veneer of glacially derived soil lies one of the signature structural features that indent the basement rocks of the North American continent. The spoon-shaped Illinois Basin covers 60,000 square miles of western Kentucky, southwestern Indiana, and central and southern Illinois. It is filled with approximately 15,000 feet of Paleozoic-age sedimentary rock, principally dolomite, limestone, and shale. These rocks harbor sixty hydrocarbon zones that have produced more than 4.5 billion barrels of oil, and as many as seventy-five Pennsylvanian-age coal seams.

Much of the knowledge of the Illinois Basin comes from analyses of the 200,000 wells drilled there since 1886 in the search for oil. Exposures of bedrock are rare and generally limited to isolated locations. One anticline, a convex-upward rock fold that bisects the bluffs along the eastern limit of the Mississippi River floodplain, in Monroe County, provides a revealing insight into the history of the basin.

The geologic genealogy of the Illinois Basin is rooted in southern Illinois. There, 600 million years ago, during Precambrian time and concurrent with the tectonic breakup of the supercontinent Rodinia, a broad-based dimple, centered upon the Rough Creek Fault Zone, began to indent the 1,500-million-year-old basement. While this riftlike structure failed to develop into a full-scale tear between continents, its location and size greatly influenced the early structural development of the basin. Thermal subsidence, caused by the cooling and contraction of gargantuan, rift-related masses of deep-seated molten rock, probably was the cause of most all of the basin development after that. The basin grew deeper as the magma cooled. As time wore on over hundreds of millions of years, the basin filled with sediments deposited by rivers and oceans.

During the Pennsylvanian period, forces related to the emergence of the ancestral Rocky Mountains, 2,000 miles to the west, periodically stressed the basin. Like the compression of the bellows of a massive accordion, these events created a series of anticlines within the sedimentary rocks filling the basin, anticlines that much later trapped migrating, subsurface streams of oil, concentrating them into economic reservoirs. By the end of the Paleozoic era, the Illinois Basin had reached a mature state of growth, ringed by a fortresslike series of buried, curvilinear and uplifted domes and arches that functioned as an identifying rim of the basin and gave it an aura of enclosure and security.

The trend of the crest (axis) of one of the better exposed of these structures, the Valmeyer anticline, extends for 12 miles between the cities of Maeystown and Valmeyer. Both sides of the anticline and its crest can be seen in cross section, west of Valmeyer, by proceeding east 0.4 mile from the intersection of Illinois 156 and the railroad tracks. At stop 1, the chert-rich exposure of limestone inclined at 15 to 33 degrees forms the southwest flank of the anticline. Another 0.5 mile to the east (stop 2) the limestone has no inclination, indicating this is the crest of the anticline. Continue another 0.7 mile and note, on the north side of the road (stop 3), strata inclinations of 3 to 5 degrees, which mark the northeast flank of the anticline—opposite in direction to the inclination observed at the first stop.

The study of this seemingly unimposing, 1.2-mile-long exposure of rock is an excellent example of geologic forensics work at its best—in this case opening the door to a better understanding of the architecture and history of the Illinois Basin and its buried treasure of petroleum and coal.

The birth of the Illinois Basin is traced to the activation of the Rough Creek Fault Zone, a riftlike rupture that formed during the breakup of the supercontinent Rodinia. Numerous anticlines (brown lines), including the Valmeyer, later developed as a result of tectonic compression related to the formation of the Rocky Mountains. When fully developed the basin was geographically accentuated by an enveloping series of arches and domes.

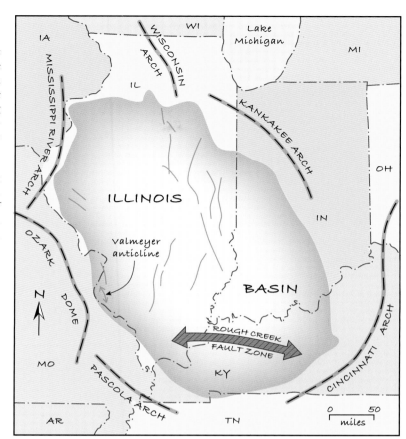

Map and cross section showing the three easily accessed exposures of the Valmeyer anticline along a 1.2-mile stretch of Illinois 156. Note the radical changes in dip angle and dip direction of the limestone and shale between stops 1 and 3.

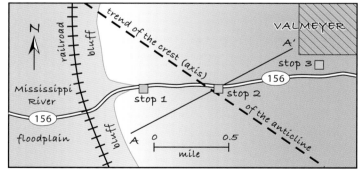

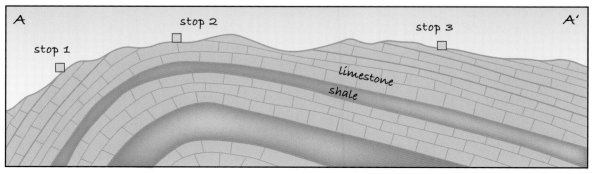

24. Hanging Rock Klint, Indiana

40° 49' 48" North, 85° 42' 26" West
Silurian Period Reef Development

One of thousands of fossil reefs that define an inland sea that once inundated most of North America.

For centuries Peruvian fishermen worried about their anchovy catch have dreaded the phenomenon known as El Niño, a weather event marked by warmer than normal temperatures in the eastern, equatorial Pacific Ocean. Its history, however, is much older—at least 130,000 years—and its victims far more significant than a seasonal catch.

The strongest recorded El Niño event occurred in 1997–1998, when drought in the rainforests of Indonesia coincided with the wettest autumn in Florida in a century. More importantly, the increase in temperature of the equatorial oceans caused an estimated 16 percent of the world's reef systems to die. Humanity can ill afford calamities such as this. The aquatic communities associated with reefs—considered the rainforests of the sea—feed a significant proportion of the world's population. Reefs have been part of the geologic scenery since the days of the Cambrian explosion, when a variety of large, multicelled life-forms suddenly appeared on the scene.

Reefs are shallow-water, wave-resistant, moundlike structures of calcium carbonate built around a rigid framework composed of the skeletal material of calcareous organisms. While corals, a marine animal related to anemones and jellyfish, are important in this construction, the term *coral reef* is something of a misnomer. Other forms of marine life, numbering in the hundreds of thousands, also contribute to these structures.

The valley of the Wabash River in Wabash and Huntington counties, Indiana, is home to some of the best-studied fossil reefs in the world. At least forty Silurian-age reefs—representative of hundreds of others that have yet to be exhumed by erosion—are exposed along the 15 miles separating the towns of Andrews and Wabash. Two of these, because of their state of preservation, are of paramount importance in understanding the architecture of these 425-million-year-old structures.

The Wabash reef, measuring 750 feet long by 40 feet high, is beautifully exposed within a railroad yard cut adjacent to the termination of East Market Street in the city of Wabash. Here, layers of nonreef rock that blend into strata of late Silurian Mississinewa shale envelop a massive altered limestone reef.

Several miles to the northeast, Hanging Rock reef, a geologic feature designated a national natural landmark, towers 100 feet above the otherwise flat, agriculture-blanketed landscape, 1.5 miles southeast of the village of Lagro. Unlike Wabash reef, which was exposed through human excavation, this is a classic example of a klint, a fossil reef exhumed by erosion and made into a prominent topographic anomaly. Although both these reefs were built through the activities of literally thousands of marine animals, fossilized animal remains are scarce, probably because of diagenesis, the physical and chemical processes that alter the nature of a rock and often destroy the fabric of any skeletal remains.

The fossil reef area of Indiana is part of the central sector of the proto–North American continent that was inundated by seas during the Silurian period, when the continent straddled the equator. Since regional mountain building forces were absent, and shallow, warm, oxygenated water conditions were prevalent, this was an ideal time and place for reef development. Indiana fossil reefs are a valued source of the high-quality aggregate rock used in the construction trades. To date, they appear to be barren of oil, but identical structures in Michigan have produced over 1 billion barrels of petroleum.

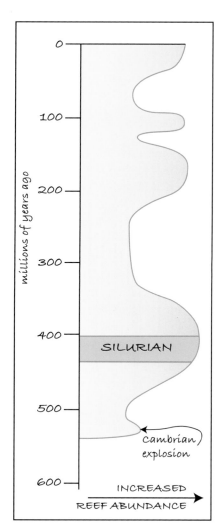

For 542 million years reef-building animals have fluctuated in number in response to changes in the environment. During the Silurian period they proliferated.

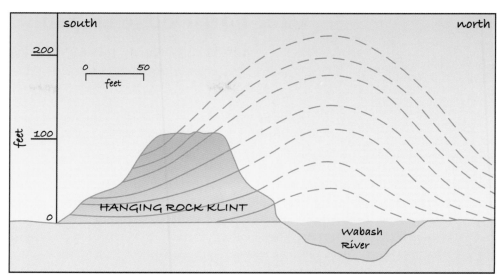

Only the southern half of the Hanging Rock klint exists today. Wabash River action has eroded the remaining portion (dashed lines).

A copse of trees forms a portal to the Hanging Rock klint.

25. Fort Dodge Gypsum, Iowa

42° 31′ 05″ North, 94° 10′ 57″ West
Jurassic Period Sedimentation

This remnant of a once vast rock layer offers clues to the climate of middle Mesozoic America.

Human history is conventionally subdivided into relative time periods, such as modern, medieval, and ancient. Historical geology—the ultimate form of chronological analysis because it engages the entire 4.6-billion-year span of Earth history—uses absolute time; for example, the Cambrian period, 542 to 488 million years ago. Whatever the choice, the purpose of assigning ages to events or objects is to weave often-disjointed facts into chronologic perspective. A dramatic case in point is a 25-foot-thick bed of gypsum that underlies a limited area of Iowa. How old is this isolated outcrop, how did it come to be, and what is its relevance to the broad scheme of Earth history?

In 1852, the Fort Dodge Formation covered 30 square miles in Webster County. Because this is one of the purest gypsum deposits on Earth, its extraction has long been economically important. At present production rates, in 2050 the final ton of gypsum will be processed into wallboard, food additives, soil conditioners, and roofing material. One of the few exposures not yet threatened by mining lies along Soldier Creek in Snell-Crawford Park, near the intersection of Williams Drive and 12th Avenue North, in the city of Fort Dodge.

Scholars have used fossil content and association with enveloping rock units to jostle the age of this gypsum from the Pennsylvanian to the Cretaceous period. More recently, the study of coniferous pollen grains has indicated the gypsum is 155 million years old, placing it in the Jurassic period, and was deposited in an environment of evaporation at latitudes ranging from 10 to 30 degrees north.

The Fort Dodge Formation is composed of three distinct members: a conglomerate at its base, a thick middle unit of 96 percent pure gypsum, and a blushing red upper sequence of sandstone and siltstone. A body of seawater in a shallow, restricted basin evaporated to form the middle unit. When the concentration of mineral salts reached a critical point, gypsum crystals formed and settled to the basin's floor. After the final drop of seawater had vaporized, the heat of the semitropical sun desiccated the exposed gypsum and created randomly oriented sets of vertical fractures that migrating groundwater later widened into deep crevices. Finally, developing river systems and meltwater from retreating Pleistocene-age glaciers deposited layers of sand and silt within and on top of the crevices.

The cycle of deposition and evaporation that created the Fort Dodge Formation took place as the result of a great inland ocean—the Sundance Sea—that covered vast regions of the midwestern United States. For decades, conventional wisdom dictated that its eastern shoreline stretched from central Kansas to North Dakota. That changed once the Fort Dodge gypsum, as well as other, similar isolated outcrops to the north in Minnesota and Canada, had been relegated to the Jurassic period, providing the evidence that cartographers needed to extend the Sundance shoreline more than 300 miles to the east. In this fashion, outcrop by outcrop, geologists rewrite the chapters of Earth history.

Interestingly, this bed of gypsum once played a leading role in one of the great hoaxes of the nineteenth century, when a New York farmer "discovered" the 10-foot tall Cardiff Giant, a "petrified man," in his plowed field. Later, archaeological scholars proved it a fake, carved surreptitiously from a massive 10,000-pound block of Fort Dodge gypsum. The popular saying "There's a sucker born every minute" is attributed to the fact that thousands of people paid hard-earned money to see this "eighth wonder of the world."

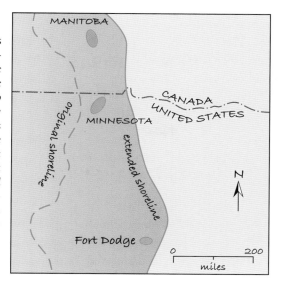

Geoscientists extended the once-accepted Sundance Sea shoreline hundreds of miles to the east when they found that rocks (green) of the same age as those at Fort Dodge also exist in Minnesota and Manitoba. (Modified after Anderson and McKay, 1999.)

An exposure of the Fort Dodge Formation upstream from the Soldier Creek footbridge in Snell-Crawford Park. The whitish rock is gypsum. —Courtesy of Ed Faires, Kernersville, NC

Deep crevices in the Fort Dodge gypsum, some more than 3 feet wide, formed prior to the deposition of younger red-bed material (background). —Courtesy of Ray Anderson, Iowa Geological Survey

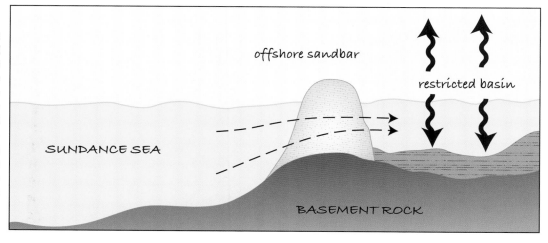

The layers of gypsum (green) formed through a process of evaporation (up arrows) and precipitation (down arrows) within a restricted basin that periodically refilled with salt water, starting the process over. (Modified after Anderson and McKay, 1999.)

26. Monument Rocks, Kansas

38° 47' 36" North, 100° 45' 45" West
Cretaceous Period Wind and Water Erosion

Lauded as the Chalk Pyramids, these sentinels demonstrate the awesome power of wind and water.

The geologic agent erosion involves the transportation of weathered rock by water, wind, and ice. Water is by far the most important of these movers and shakers. Globally, its volume exceeds 332 million cubic miles, 97 percent of it in oceans and seas. The remainder, in the form of glaciers, groundwater, lakes, and rivers, sculpts the land. As the largest river in the United States, the Mississippi annually transports as much as 500 million tons of sediment to the Gulf of Mexico.

Overall, wind erosion is not as serious a problem, but in regions with minimal precipitation it becomes significant. For example, during one brief storm in California's San Joaquin Valley, 200-mile-per-hour winds eroded an estimated 100 million tons of soil. The long-term effect of a combination of water and air in motion can be dramatic. An excellent example is Monument Rocks National Natural Landmark of Gove County, Kansas, where the Monument Rocks, also known as the Chalk Pyramids, stand as vestigial evidence of an inland ocean that once inundated central North America.

Ninety million years ago, the North American Plate slowly converged with the Pacific Plate, causing the central area of the continent to sag. Waters from the Arctic Ocean and the Gulf of Mexico gradually coalesced within the newborn trough. This Western Interior Seaway extended 600 miles from the embryonic Rockies east to the Appalachians, and at least 3,000 miles north to south. Geologists describe this as one of the greatest epicontinental seas of all time, on average 500 feet deep and characterized by a placid, tropical environment.

As the Cretaceous period closed, the growth of the Rocky Mountains elevated the land surface of North America and forced the Western Interior Seaway to retreat. In its wake lay a vast, relatively thick sequence of sedimentary rock composed predominately of shale, chalky limestone, and chalk. Within the heart of this rock series lies the majestic 550-foot-thick Niobrara Formation, parent to the Monument Rocks.

For 65 million years, the Niobrara has experienced a steady siege of erosion, initially by streams, and then through the action of wind. Today, only isolated islands—the chalk-rich Monument Rocks—remain. Chalk is an unusual type of rock. It is porous, relatively soft, fine textured, earthy, usually white to gray, easily whittled, and composed almost entirely of calcite. A hand lens reveals multitudes of extremely small, pinhead-sized marine fossils—chiefly foraminifera, a class of single-celled, floating organisms—plus the remains of calcareous algae. The Niobrara Formation, 95 percent of which is chalk of this very nature, is quite similar in composition to the famed White Cliffs of Dover of southern England.

The Monument Rocks also contain a fascinating array of large mosasaur, ichthyosaur, and plesiosaur fossils—all former predatory marine reptiles, the latter the inspiration for the fabled Loch Ness Monster of Scotland. Numerous fish fossils are also present, the most famous being the "fish within a fish," which is on display at the Sternberg Museum of Natural History in Hays, Kansas. The fossil comprises a perfectly preserved, 14-foot-long *Xiphactinus* with a 6-foot *Gillicus*, a tarponlike victim, enclosed within its rib cage. Eighty-three million years ago this predator—the largest bony fish that ever lived—died soon after swallowing its prey, a fatal act of biting off more than it could chew.

The Western Interior Seaway divided the continent to an extent not seen since.

Pioneers traveling the Butterfield Overland Stage Route used Monument Rocks as a trail marker. The erosive forces of wind and water that shaped the Keyhole will continue until its roof collapses, leaving a pair of spires that bookend a distant sibling.

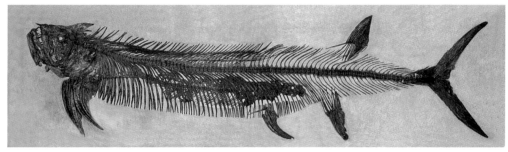

The fish within a fish fossil, said to be the most widely photographed fossil in the world, preserves a defining moment of gluttony. —
Courtesy of the Sternberg Museum of Natural History, Hays, KS

27. Ohio Black Shale, Kentucky

38° 09' 59" North, 83° 35' 31" West
Devonian Period Sedimentation

The largest natural gas field in the Appalachian Basin is sourced from this coal-black, carbon-rich rock.

Of the three readily recognized categories of sedimentary rock—sandstone, mudstone, and limestone—which is most common? Answer: More than half of all sedimentary rock is mudstone, a nonlayered rock composed of grains so small—less than 0.004 millimeter in diameter—that 1 ounce may contain thirty to sixty billion individual particles.

The bedrock of the United States is unusually rich in mudstone, especially the variety that is easily split into thin plates, which is known as shale. One particular shale formation has garnered a lot of publicity because of its importance as a hydrocarbon resource. Deposited in an epicontinental sea that covered most of the proto–North American continent during Middle and Late Devonian time, this shale goes by several names—Chattanooga, New Albany, Cleveland, and Antrim—but to many Midwestern geologists the name Ohio Black Shale rings the bell. The entire formation is on display in the 200-foot-high roadcut at mile marker 130.3 along I-64, about 49 miles east of Lexington. The predominant black color derives from organic matter, which in some strata exceeds 8 percent by weight, that gives the Ohio Black Shale its lofty economic status as a prime source of natural gas.

During the Devonian period, proto–North America was 10 to 20 degrees south of the equator. Dominating its topographic profile were the Acadian Mountains, the second stage in the long-term growth of the Appalachian Mountains. The mountains were created by the collision of proto-Europe with proto–Scandinavia and Asia. The vast, shallow Kaskaskia Sea lay to the west, its surface waters minimally ruffled because the mountains sheltered it from prevailing winds. The sea's high biologic productivity showered the seafloor with abundant dead organic matter that depleted the deep water of any lingering dissolved oxygen, creating a seafloor environment that minimized decay and maximized preservation. As sedimentation buried this organic matter, pressure and temperature altered it, first to petroleum and finally to natural gas. The stage was thus set for the Ohio Black Shale, millions of years later, to become a top-drawer gas producer. And indeed it has.

Approximately three-quarters of the natural gas produced annually from the Ohio Black Shale in Kentucky comes from the Big Sandy Gas Field, which at 2,344 square miles ranks as the largest natural gas field in the Appalachian Basin. Since its discovery in 1892, each of its approximately three thousand wells have averaged a total production of 300 million cubic feet of gas, enough to keep the typical American home warm and fuzzy for 3,500 years.

Should this shale ever run out of gas, it could well maintain its celebrity status based on its remarkable assemblage of fossils. During the Devonian period, also known as the Age of Fishes, when marine vertebrates dominated the saltwater world, a pioneer form of lungfish called crossopterygian appeared. Paleontologists deduce that amphibians—the first vertebrates able to gulp air—evolved from this early air breather. The fossil bone and fish plate (skin) remains of twenty-two species of these extinct, armor-headed fish and impressions of *Foerstia*, a form of seaweed—all found in the Ohio Black Shale—give testimony to this extraordinary time of evolution. With life emerging from the sea, Earth would never again be the same.

Multiple beds of black to gray mudstone and shale, with fossil imprints of fish scales, fish teeth, and an early form of algae, define the exposure at mile marker 130.3.

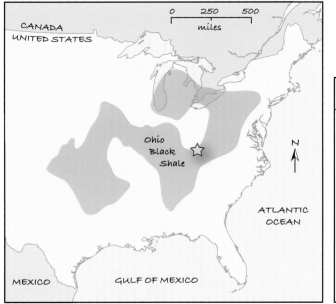

The Ohio Black Shale covers a large sector of the Midwest. The star marks the roadcut at this geo-site.

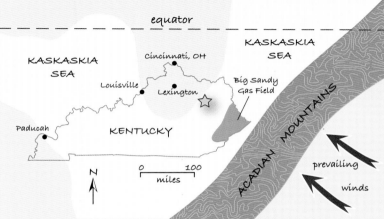

The Kaskaskia Sea lay to the west of the Acadian Mountains in Devonian time. Note the position of the equator. Some wells in the Kentucky portion of the Big Sandy Gas Field have produced natural gas for fifty years. The star marks the roadcut.

73

28. Mammoth Cave, Kentucky

37° 11' 12" North, 86° 06' 04" West
Neogene Period Cave Formation

This longest cave system in the world is a 390-mile-long, multilevel scene of dissolution and beauty.

Caves and carbonate rock go together like baseball and a bratwurst sandwich. One of the signature aspects of carbonate rock is its propensity to be dissolved by acidic solutions. Since both carbonate rock and acidic solutions abound in nature, caves also abound. Nowhere in North America are caves more prevalent than south-central Kentucky, also know as the "land of 10,000 sinks."

Carbonate rock is a by-product of shallow, sunlit seas that teem with marine life, such as algae, coral, mollusks, and foraminifera. In life, these animals extract calcium carbonate from the environment to form external skeletons, their shells. In death, their skeletal material settles to the seafloor as great thicknesses of calcareous ooze. With time and pressure, the ooze hardens into limestone, a common sedimentary rock that is easily dissolved.

South-central Kentucky is built on a solid base of limestone, deposited within an epicontinental sea that inundated the southeastern United States during the early Carboniferous period. Over the last 20 million years, groundwater charged with carbonic acid—the same acid found in soft drinks—has slithered through cracks and fissures in this bedrock. The interplay of chemistry and dissolution—the process of dissolving rock—systematically enlarged the microscopic passages into ever-larger voids that eventually aggregated to today's cave systems. This same process will eventually destroy this wonderland, because limestone-based, subterranean drainage systems, like their surface cousins, are ephemeral. In geology, as in life, the only constant is change.

Today, south-central Kentucky is one of the best-developed karst landscapes known on Earth. Characterized by sinkholes and underground drainage, the terrain is a maze of tunnels, amphitheaters, and passageways. Estimated to stretch almost 1,000 miles, the labyrinth of Mammoth Cave is the world's longest known subterranean cavity system. Mammoth's fame as a goliath cave largely derives from how its principle passageways are positioned. Like a multistory building, they are stacked one upon the other. Unlike a building, however, the upper level of Mammoth Cave was the first to develop and the lowest level the latest. Why? As the Green River—the master surface stream of the region—erodes its valley ever deeper, the groundwater table (highest elevation of ground saturated with water) drops ever lower, allowing the development of passageways in the overlying mass of limestone. The primary route for migrating groundwater lies along the major bedding planes of the limestone formations. Dissolution enlarged these primary routes to form hallways varying from single-file width to auditorium-sized.

Wherever water cascades downward along fractures, vertical "elevator shafts" form, connecting one level with another. These shafts are perhaps the most breathtaking features of Mammoth Cave, from the smallest youngster to the granddaddy of them all—the 192-foot-high Mammoth Dome. Flowstone, a thin, sheetlike calcareous deposit, embellishes the rock area adjacent to the 138 steps imbedded in the wall of this oval-shaped shaft. Eons of dripping water deposited the flowstone.

Interestingly, stalactites and stalagmites are uncommon, because the impermeable sandstone that functions as a cap-rock limits how much water can enter and flow through the limestone beneath it. Deposited by an ancient river during the Pennsylvanian period, the sandstone has served to ensure the almost pristine nature of what is, to all intents and purposes, a magnificent underground cathedral.

Three different limestone formations are exposed within the irregular walls of Mammoth Dome.
—Courtesy of the National Park Service

Exposure of cave-forming limestone along the road leading into Mammoth Cave National Park.

Ancient river currents periodically changed direction and created crossbedded sandstone that today keeps the underlying limestone in the Mammoth Cave region from dissolving. Caves can form beneath the surface where this caprock has been eroded.

29. Four Corners Roadcut, Kentucky

37° 17' 21" North, 83° 13' 15" West
Pennsylvanian Period Coal Formation

A 3-D exposure of combustible rock exemplifies the geographic heart of a huge hoard of black gold.

Media reports of America's fossil fuel crisis emphasize that close to one-half of the petroleum the United States uses on a daily basis is imported. A significant portion comes from Saudi Arabia, the desert kingdom with 20 percent of the world's proven oil reserves—more than any other country. An inconvenient thought, yes, but is *petroleum* a synonym for *fossil fuel*?

Think *coal*—that other naturally occurring, carbon- and hydrogen-based substance, and the outlook regarding the crisis brightens significantly. Earth contains some 1,000 billion tons of economically recoverable coal, 27 percent of it in the contiguous U.S. states—more than in any other country. If Saudi Arabia is the powerhouse in the world of oil, then the United States is the Saudi Arabia in the world of coal.

Coal is a combustible, organic, sedimentary rock composed primarily of hydrogen, carbon, and oxygen and derived from the remains of mats of plant life that once flourished in a tropical swamp. The combination of heat, pressure from overlying sediment, and time measured in millions of years alters these mats into rock that falls somewhere in the spectrum of the coal series. This sequence delineates incrementally higher levels of fixed carbon—the percentage of solid combustible matter in a rock after the removal of moisture, ash, and volatile matter—that ranges from peat, with a carbon content of 60 percent, to lignite, bituminous, and finally anthracite, with a carbon content of 95 percent. The greater the heat, pressure, and time, the greater the amount of fixed carbon. Coal formation in the United States began on a large scale during the Pennsylvanian period, when conditions, ranging from marsh to lake, delta, and shallow marine waters, were ideal for extensive plant growth.

The Appalachian Basin, extending from Pennsylvania to Alabama, is one of the most important and historic coal producing regions of the world. Formed by the coalescence of the landmasses of Laurasia and Gondwanaland, this gargantuan crease within Earth's crust dates back in time to the Taconic orogeny, the first of three mountain building episodes that ultimately formed the modern-day Appalachian Mountains. Of the estimated 35 billion tons of bituminous coal mined from the Appalachian Basin over the past three hundred years, two-thirds has come from the central sector, which extends from northern Tennessee to central West Virginia. The Bluegrass State portion of the central sector, called the Eastern Kentucky Coal Field, encompasses thirty-seven counties and includes approximately ninety beds of coal, ranging from 1 to more than 9 feet in thickness. Hundreds of mines, all closed to the public, dot the region.

At the Four Corners roadcut, near the intersection of Kentucky 15, Kentucky 80, and the Hal Rogers Parkway, 2 miles north of Hazard, several seams within the 311-to-307-million-year-old Breathitt Formation are exposed in magnificent clarity. Three "black gold" beds are on display: the 7-foot-thick Hazard No. 7, the 8.5-foot-thick, economically important Hazard No. 8, and the 3-foot-thick Hazard No. 9. The shale and siltstone units separating the No. 8 seam from the No. 7 seam represent a forest environment, as evidenced by the fossil remains of more than two dozen stumps and tree trunks, some measuring 3 feet in diameter. Many plant fossils and a few brackish-water fauna fossils also can be found within the exposed shale strata of this roadcut, representing sediment deposited in grassland and coastal swamp environments.

These seams, like the coal being mined throughout the Eastern Kentucky Coal Field, typify the fossil fuel that made nineteenth- and twentieth-century America an industrial giant. They likely will remain an invaluable resource well into the future.

This sample of Hazard No. 8 coal typifies the coal beds that make up the vast reserves of the Breathitt Formation of eastern Kentucky. Keys for scale.

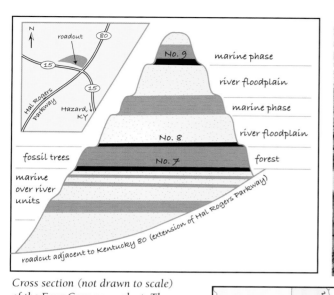

Cross section (not drawn to scale) of the Four Corners roadcut. The coal seams (black lines) formed in a swamp environment. (Modified after Chesnut et al., 1986.)

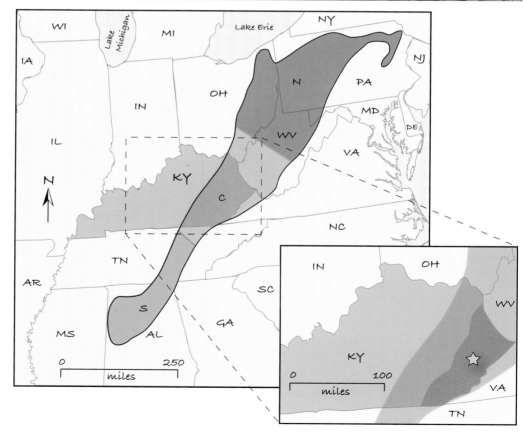

The Appalachian Basin (outlined in black), the most productive of all U.S. coal basins, comprises three coal-bearing sectors: southern (S), central (C), and northern (N). The Four Corners roadcut (star) lies in the center of the Eastern Kentucky Coal Field (dark orange, inset), which is contained in the central sector. (Modified after Ruppert et al., 2002.)

30. Avery Island, Louisiana

29° 53' 46" North, 91° 54' 25" West
Jurassic Period Salt Intrusion

A miles-high plug of salt that is but one of a forest of similar columns underlying the Gulf of Mexico.

For more than 100 million years the muddy waters of the Mississippi River system have transported rich, virgin soil away from its Midwestern origins. This one-way, southbound drainage ends in the swampy coastlands of Louisiana, where, on average, more than 1 million tons of sediment are deposited daily into the Gulf of Mexico. A well in downtown New Orleans would have to be 40,000 feet deep to reach the base of this geologically young pile of sand and mud. This is the bayou and salt marsh world of southern Louisiana, the land of Henry Longfellow's Evangeline and the site of one of the richest hydrocarbon provinces in the world.

Strip away this coating of deltaic sediment and a new world emerges—one of a "forest" of colossal, spiny columns of rock soaring upward from a thick parent unit of salt that extends from Texas to Alabama. More than six hundred so-called salt domes—structural intrusions of sodium chloride—comprise this forest.

The mother bed common to all these structures, named the Louann Salt, dates to the Jurassic period. The youthful Gulf of Mexico, created by the breakup of the supercontinent Pangaea, was at that time periodically isolated from the open ocean. The evaporation of ponded saline waters produced the 1,500-to-12,000-foot-thick bed of salt that was subsequently covered by numerous layers of Mississippi River sediment. As the salt was buried deeper, it began to flow upward in a plastic fashion. Less dense than the overlying sedimentary rock and pressurized by the sedimentary overburden, it rose until it reached density parity, when its density matched the overlying and surrounding sediments. Many of the domes have mushroom-shaped crowns, a sign of advanced development. Others, appearing more like tall mounds than crowned columns, are deemed to be younger.

Along the coast of Louisiana an unusual alignment of mounds, known as the Five Islands, parallels US 90 between Morgan City and New Iberia. The core of each is a column of salt, arching rather than piercing the land surface. The most famous of these salt domes is Avery Island, located southwest of New Iberia at the termination of Louisiana 329. Here, a 45,000-foot-tall shaft—nearly 8.5 miles high—of 99.1 percent pure salt penetrates to within 8 feet of the island's domed surface, which is 163 feet above sea level at its highest point. Industrial-grade salt has been mined since 1898 by the room-and-pillar method, which is the excavation of roomlike openings leaving intervening salt pillars to support the roof of the mine. Presently, miners are extracting salt from the 1,600-foot level of this 1.5-to-2-mile-diameter structure, estimated to contain 150 billion tons of salt.

The nature of the upward flow of salt, which began millions of years ago and continues at approximately 1 millimeter per year, is evidenced by patterns of slightly contorted bands that compose the mine walls of the Avery Island dome. Salt under pressure and in the presence of heat moves very much like the material in a lava lamp, slowly swirling its way upward. Oil and gas also play a role in the economy of Avery Island. The impervious salt abuts layers of sandstone and shale. With shale being the source rock of hydrocarbon and sandstone an ideal reservoir rock—a porous and permeable rock that contains oil or gas—a well drilled parallel to the edge of the salt pillar can potentially penetrate stacked concentrations of oil or natural gas. The salt and oil sectors of Avery Island remain private, but both the 250-acre botanical garden and the on-site Tabasco bottling operations are open daily to the public.

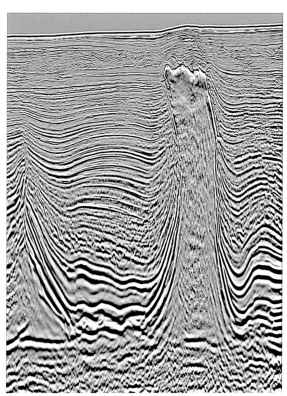

High-tech depiction of a 4.3-mile-high salt dome intruding formations that underlie the Gulf of Mexico. —Courtesy of TGS-NOPEC Geophysical Company, Houston, TX

Translucent salt specimen from the 1,600-foot depth at Avery Island. Pen for scale. —Courtesy of Charles Herron, Wilmington, NC

Cross section of a salt dome showing hydrocarbon accumulations (blue) within beds of sandstone (orange). Note the mushroom-shaped crown, indicating an advanced state of development.

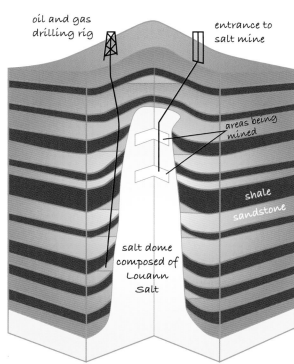

Irregular layers are evidence of upward salt motion. As it moved, the salt incorporated chunks of sandstone (red blob in center) into its mass from adjacent sedimentary rock.

31. Schoodic Point, Maine

44° 19' 57" North, 68° 03' 40" West
Devonian Period Mountain Building

Drifting continents and colossal collisions resulted in a dike-infested coast of majestic form and beauty.

The southeast coast of Maine, the Down East region traveled by generations of captains sailing out of Boston, is a wonderment of topographic indentation defined by numerous bays, coves, and craggy estuaries. This land of metamorphic and igneous rock was born of tumultuous continental impact and scorching volcanic activity—processes that disallow the presence of sedimentary strata. From Penobscot Bay to the New Brunswick border these rocks tell a tale of their distant offshore conception, transoceanic migration, and ultimate arrival in the New World. The whole of this account is conveniently contained within the borders of Acadia National Park, reached by Maine 3 south from Ellsworth. The early paragraphs are recorded there on Mount Desert Island, which is the principal area of the park. The concluding statements are enshrined within the wave-splashed granites of Schoodic Point, several miles to the northeast.

As the curtain was rising on the Cambrian period, a small chunk of continent—Avalonia—broke away from Gondwanaland, the landmass that would someday fracture into South America, Australia, Africa, and Antarctica. For 120 million years this minicontinent "sprinted" to the north and west, bound on a collision course with Laurentia, or proto–North America. Being larger, Laurentia was "plodding" westward at an estimated 1 inch per year, a third the speed of the more agile Avalonia.

By Middle Devonian time, the smaller continent had docked with Laurentia in a collision that reached from Maritime Canada south to the Carolinas. This clash of landmasses sent a massive shiver of metamorphism throughout both and compressed and heated their overcoat of mud, sand, and volcanic ash to formations of slate and schist, which today are exposed along the north and south shores of Mount Desert Island, enveloping massive domains of erosion-resistant granite.

The rest of this story is scenically exposed at Schoodic Point, south of the hamlet of Winter Harbor. When the leading edge of Laurentia plunged beneath Avalonia, the resulting friction created temperatures hot enough to melt rock, causing great cauldrons of acidic magma to rise to the surface. Some of it extruded onto the surface, while the rest solidified underground, guaranteeing that this future shoreline of Maine would long possess a rock-solid foundation of granite.

As the acidic magma bodies cooled, they contracted and cracked. New injections of basaltic, dark-toned magma filled in the fractures and crystallized as dikes—tabular igneous intrusions that cut across the preexisting rock. More than once, a generation of younger dikes cut across older dikes, and these intersections are clearly visible at Schoodic Point. Other dikes at this location taper over a distance of several feet, gradually fingering out, or declining to zero thickness. And at one picture-perfect location, storm and tide action has completely eroded away part of a massive dike, leaving a vertical-walled chasm that echoes the thunder of crashing waves. The geologic interest and the scenic beauty of Schoodic Point are primarily associated with the contrast between these black dikes and the pinkish-hued granite.

When this tumultuous period of igneous and metamorphic activity climaxed, the Acadian Mountains, created by the second stage of tectonic activity that ultimately formed the Appalachian Mountains, pierced the clouds to elevations perhaps as lofty as the present-day Himalayas. Even though erosion has long since eliminated that former scene of grandeur, Mount Desert Island and Schoodic Point remain weather-beaten examples of the geologic carnage that occurs when landmasses collide in the plate tectonic process that continues to rearrange the terrestrial and oceanic geography of Earth.

Fine-grained dike intruding coarse-grained granite at Schoodic Point. —Courtesy of Henry N. Berry IV, Maine Geological Survey

Rapid heat exchange at the contact between invading basalt magma (dark) and cold granite bedrock (light colored) created flow lines, mineral streaks that indicate the direction the magma was flowing before it crystallized. Dime for scale.
—Courtesy of Henry N. Berry IV, Maine Geological Survey

Schoodic Point dikes weather and erode at a faster rate than the granite country rock because of their mineral make-up and degree of fracturing. —Courtesy of Henry N. Berry IV, Maine Geological Survey

32. Calvert Cliffs, Maryland

38° 40′ 45″ North, 76° 31′ 58″ West
Miocene Epoch Fossilization

Perhaps the very best site in the United States to collect a handful of shark teeth in a matter of hours.

Geologists are historians of an unusual stripe, with an almost addictive craving to stamp a date on the many events that lead to a better understanding of Earth. Before the use of radioactivity to define geologic time in absolute terms, the easiest way to date rock was by applying one of two principles that establish relative age. The principle of superposition holds that in a sequence of undisturbed strata the youngest lies on top and the oldest at the bottom. The principle of faunal succession states that with time newer organic forms and species replace older ones. The cliffs of Bayfront Park, also known as Brownies Beach, immediately south of Chesapeake Beach and off Maryland 261 in Calvert County, Maryland, are an ideal starting place to better understand how these two concepts help determine which rock is older and which younger.

The shoreline exposures in Bayfront Park are only one small part of the Calvert Cliffs, the wave-cut shoreline that extends some 30 miles along the eastern edge of the Calvert County portion of Chesapeake Bay, the largest estuary in North America. The cliffs are composed of three formations of sandstone and shale that comprise the surface geology of this area. The Calvert Formation is exposed in the north in the vicinity of Bayfront Park, followed to the south by the Choptank Formation, and concluding with exposures of the St. Marys Formation, which extends beyond the community of Cove Point. Because these rocks dip to the southeast, at a very low angle to the horizontal, it is obvious that both the Calvert and Choptank formations underlie the St. Marys Formation. Employing the concept of superposition, the St. Marys must therefore be the youngest of the three. With that understood, how does the application of faunal succession apply to the relative age of these rocks? Answer: Geologists have subdivided these formations into twenty-four zones, each with a unique suite of fossils. If an assemblage of fossils identical to any one of these were found in rocks many miles from Calvert County, the law of faunal succession would suggest the two suites of rocks were the same age. In these two manners of interpretation, geologists around the world have approached the task of assigning relative ages to rock bodies.

The Calvert County formations were deposited in warm, highly oxygenated marine waters bordered by a cypress swamp. It was home to a variety of life: seals, whales, stingrays, and porpoises, plus a myriad of invertebrates such as oysters, clams, and snails. Not surprisingly, the cliffs today are known as the most significant fossil-bearing deposit of Miocene-age rock exposed along the entire eastern seaboard of the United States. Even more important, this environment attracted an amazing variety of predatory sharks. Because one shark can shed up to 1,800 teeth in a year—and thus tens of thousands of teeth in a lifetime—the Calvert Cliffs region is a world-class mecca for collecting fossil shark teeth. Fossil hunters can find teeth from as many as eighteen different species.

Fossil teeth constantly erode out of these semiconsolidated rocks. Within only an hour or so even amateur collectors can gather from the shoreline sands at Bayfront Park a handful of variously shaped teeth from different species. They range in size from minute to the much sought after and rare, 7-inch-long teeth of the megalodon—classed as the "king of jaws." The most common teeth in the Calvert Cliffs come from gray (45 percent), tiger (20 percent), snaggletooth (15 percent), sand (5 percent), and mako sharks (4 percent). To the inquiring and imaginative mind this particular stretch of beach offers a personal and up close peek into two worlds: one the nature of scientific reasoning, and the other a daunting world filled with hunters and the hunted.

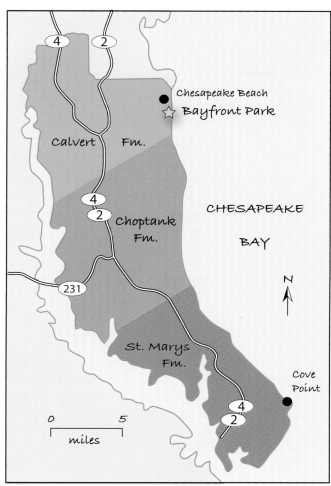

Geologic map showing the formations of Calvert County.

A combination of persistence, a sieve, and a trowel spell success in finding fossil teeth in the beach sands. Digging in the cliffs is not recommended as they are quite prone to collapse.

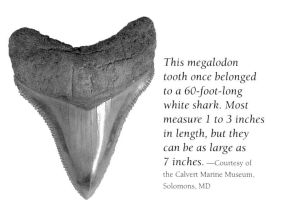

This megalodon tooth once belonged to a 60-foot-long white shark. Most measure 1 to 3 inches in length, but they can be as large as 7 inches. —Courtesy of the Calvert Marine Museum, Solomons, MD

Shark teeth commonly found at Bayfront Park include (A) mako, 2 inches long; (B) gray, 0.375 inch; (C) tiger, 0.5 to 0.75 inch; and (D) snaggletooth, up to 1.5 inches. —Courtesy of the Calvert Marine Museum, Solomons, MD

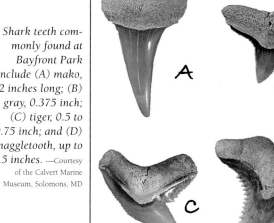

33. Purgatory Chasm, Massachusetts

42° 07' 44" North, 71° 42' 54" West
Pleistocene Epoch Cold Weather Erosion

A disordered, commingled, and chaotic setting of rocks with a mysterious origin.

In science, hypotheses lead to theories, and theories to laws, with the entire process marked by trial and error and the cross-fertilization of ideas. "Facts" remain ever subject to revision based on new discoveries. Over several centuries, thousands of educated experts—chemists, physicists, biologists, and geologists—have laboriously gathered the record of Earth's evolution, each studying an area of personal interest and painstakingly subjecting it to the tenets of science. Sometimes this process is easy, but often it is drawn out and controversial. A case in point is the geology of Purgatory Chasm, a state reservation located off Massachusetts 146 (exit 6) south of Worcester. After a century and half of analyses, consensus regarding the chasm's story remains elusive.

The geographic setting is easy enough to understand: a vertical cleft 0.25 mile long and 70 feet high by 50 feet wide dissects a bastion of heavily fractured granite rising from a floor littered with a jumble of angular blocks of the same composition. Sometimes called New England's "grand canyon," the site is a showcase of tortured topography. A long look raises the inevitable question: how did this chaotic scene come to be?

Scientists have advanced an array of possible causes: river erosion, shoreline abrasion, volcanism, earthquakes, and glaciation. The angularity of the boulders precludes both shoreline and river erosion, because water in motion rapidly—in the sense of geologic time—rounds rock corners and edges. Scratch these agents as the cause.

Volcanism also seems unlikely. While volcanic cones and lava flows once prevailed here, the eruptive forces creating these landforms have been dormant for at least 250 million years, essentially since the long-term tectonic pulses associated with the assembly of the Appalachian Mountains subsided. Over such a great span of time, even the most erosion-resistant rock type would have weathered to mere grains of sand and clay. Indeed, the ubiquitous presence of large angular boulders argues for recent, rather than ancient, causes. Scratch volcanism.

The two remaining possible culprits fit the need for recent cause. Earthquakes have rattled New England for millions of years—and still do periodically. As recently as 14,000 years ago the last continental glacier retreated from the region. The striations—linear scratches caused by either earthquakes or ice movement—that embellish the surface of some of the Purgatory boulders support one or both of these causes.

Consider earthquakes and glacial activity with an open mind and scientific logic, and one interesting scenario advances to the forefront: Initially, one or a series of earthquakes shook and fractured the Purgatory Chasm granite terrain. Then massive ice sheets moving across the area invaded one of these fractures, widening and deepening it until it reached the distinctive dimensions of today's chasm. And finally, water repeatedly freezing and thawing in the fractured granite—a process called frost wedging—caused the granite to break into the angular boulders that litter the chasm floor.

Another ice-related theory suggests the sudden release of dammed-up glacial meltwater near the end of the last ice age, some 14,000 years ago, formed the chasm. However, the presence of regional topographic highs that could have protected the chasm from the deleterious effects of fast-moving meltwater, and the lack of evidence that flowing water smoothed the chasm walls are strikes against this theory. With the finding of new evidence, other solutions, of course, are possible. Until then, aspects of geologic mystery and scientific intrigue continue to invade the shadowy corridors of Purgatory Chasm.

The youthful disorder of Purgatory Chasm contrasts with the mature topography of the surrounding countryside.
—Courtesy of Renee Weihn, Purgatory Chasm State Reservation

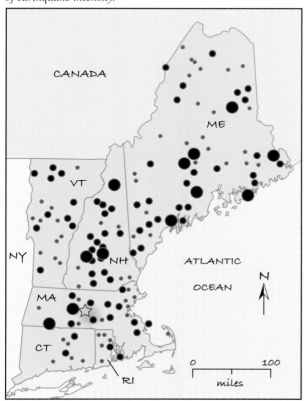

Map showing the distribution of earthquakes—a possible cause of the development of Purgatory Chasm (star)—that have shaken New England since 1975. Dot size equates to degree of earthquake intensity.

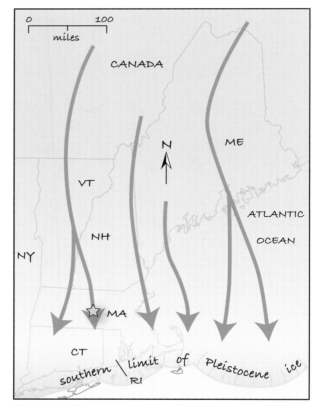

Massive ice sheets repeatedly invaded New England during the Pleistocene epoch. One of these invasions may have formed the rock rubble ruin of Purgatory Chasm (star).

34. Nonesuch Potholes, Michigan

46° 42′ 31″ North, 89° 58′ 24″ West
Holocene Epoch River Erosion

Seven-foot-deep cylindrical cavities are evidence of the power of pebbles and swirling stream eddies.

The south shore of Lake Superior is a mecca for anyone wishing to tread different paths and view untrampled vistas. Its most prominent geographic feature is the Keweenaw Peninsula, a shark fin–shaped promontory that projects from the American side of the lake. For more than a century this spine of Precambrian volcanic rock vibrated with a lumbering and mining economy. Once a full 80 percent of America's copper production occurred in Keweenaw Peninsula rocks. Today, the forests are undergoing secondary growth and the mines lie abandoned and flooded.

Notwithstanding this decline in activity, geologic analysis of the region continues. While immense quantities of copper remain entrapped in the igneous bedrock, near-term economic promise is focused on a regional sedimentary rock that was born of unusual circumstances.

One billion years ago, the Lake Superior region experienced extended tectonic assault. This was the era of the Midcontinent Rift. For 22 million years, thermal forces attempting to subdivide proto–North America stretched and tore the continent's crust in a manner much like what is happening today along the axis of the East African Rift. When these forces became quiescent, extension ceased and sedimentary rock covered the fractured wound. One of these band-aid units is the Nonesuch shale, 600 feet of siltstone and shale that was deposited within a freshwater lake environment.

A classic exposure of the Nonesuch occurs at the mouth of the Presque Isle River, north of the community of Wakefield, in the western part of the Porcupine Mountains Wilderness State Park. Here, a field of potholes overlies an area with potential for production of both Precambrian copper and petroleum. Visible only from the swinging-bridge crossing of the Presque Isle, the potholes are up to 7 feet deep, classic cases of abrasion. Caught in circular motion by eddies, pebbles abraded and scoured the bedrock and created pits of various sizes, though most are deeper than they are wide. A thick carpet of pebbles on the floor of these 3- to 4-foot-wide features conceals their real depth.

The black color of the Nonesuch comes from its organic content. Although small in volume—usually less than 2 percent—this is the very stuff that generated the crude oil sampled from a nearby copper mine. This subterranean pocket of petroleum, indirectly age dated as 1,047 million years old, is possibly the oldest hydrocarbon discovered to date anywhere in the world. A massive search for Precambrian Nonesuch oil conducted in the 1980s throughout the 1,000-mile-long Midcontinent Rift ended in failure, but the potential is still there.

More recent evaluation focused on the Nonesuch shale's metallic resources. Copper, in both native form (an element found in an uncombined chemical state) and as disseminated copper sulfide (where the desired mineral is scattered throughout the host rock), exists in amounts exceeding 50 pounds per ton of rock—at present not an encouraging volume, but nevertheless of potential future value.

The potholes of the Nonesuch shale are remarkable evidence of how stream erosion can slowly reshape the surface of Earth. In addition to being visually appealing, they mask an underlying potential of immense significance—in this case new reserves of copper and petroleum, two of the natural resources that keep the economies of American industry humming.

This pothole, which is several feet wide, is typical of the many pitting the streambed at the mouth of the Presque Isle River. —Courtesy of the Michigan Department of Natural Resources

In the Nonesuch shale, disseminated copper oxidizes to a greenish sheen (left) while the native form of copper retains its bright-penny appearance. Six-inch ruler for scale. —Courtesy of Charles Herron, Wilmington, NC

Sample of ripple-marked Nonesuch shale alongside a flask of billion-year-old of Nonesuch crude oil. Ripple marks are generally considered evidence that sediment was deposited in shallow water, where the gentle currents reworked the sediment. Six-inch ruler for scale. —Courtesy of Charles Herron, Wilmington, NC

35. Quincy Mine, Michigan

47° 08' 12" North, 88° 34' 30" West
Proterozoic Eon Mineralization

This geologic touchstone to Copper Country is the most legendary of the Great Lakes landscapes.

Found associated with more than 160 minerals, copper is a ductile, malleable, reddish brown, nonferrous metal that is heavily sought after during times of industrial and commercial expansion. In only one form, called native copper, does it appear in an unadorned state—meaning not bound with other elements. The now-dormant Lake Superior copper region has long been recognized as the world's largest native-copper mining district.

From 1845 to 1968, some one hundred mining companies supplied the economic lifeblood of the Copper Country, producing more than 10 billion pounds of refined product from a 2-to-4-mile-wide, 26-mile-long belt of mineralized lava flows. Throughout most of these years, the defining operation was the Quincy Mine, affectionately nicknamed Old Reliable because it provided ample dividends for more than three generations of investors. Open to the public, the Quincy is on the eastern outskirts of Hancock, Michigan, alongside US 41. Tours include an inspection of the world's largest steam hoist; extensive mineral and historical exhibits; and a hard-hat and overcoat excursion, via a cog-rail tram ride, into a portion of the underground mine as it might have appeared on the final day of operation.

The Portage Lake Volcanics, a series of igneous and sedimentary rocks totaling some 10,000 to 15,000 feet in thickness, crosses the Keweenaw Peninsula. It incorporates at least two hundred individual lava flows, twenty-five of them separated by relatively thin beds of conglomerate deposited by rivers during times of volcanic quiescence. These flows average 1 billion years in age, and one unit—the Greenstone flow—has a maximum thickness of 1,300 feet.

Each lava flow is composed of two merging parts. The thicker, underlying portion, which miners call "trap," is gray to black, dense, massive rock barren of minerals. In contrast, a zone of minute to large vesicles, cavities formed by gas bubbles that were trapped as the lava solidified, characterizes the top of each flow. This spongelike cap (in texture only) is described as vesicular or, if minerals fill the vesicles, as amygdaloidal. The Quincy Mine is in a narrow zone of prime, vesicle-filled flows known as the Pewabic amygdaloid, which averages 2 percent native copper by weight.

Like all the lava units of Copper Country, the Pewabic amygdaloid spewed onto the surface during the middle of the Proterozoic eon, when the proto–North American continent was being subjected to subsurface processes intent on subdividing it. These pull-apart forces, like those occurring today in Iceland and East Africa, ultimately failed, but they left the lava sequence strongly folded and faulted. At the Quincy Mine, the Pewabic amygdaloid plunges to the northwest at a 54-degree angle, its outcrop oriented along a northeast-to-southwest trace. Deep-seated, metal-bearing, hydrothermal solutions rose toward the surface of the formation and deposited native copper within its vesicles.

The Quincy no. 2 shaft, one of ten that were excavated, eventually reached an inclined depth of 1.75 miles, along which the mine was divided into ninety-two levels of operation. Today visitors explore a 2,400-foot section of the seventh level, in which twenty-six lava flows are exposed. Water floods the remaining eighty-five levels. Even though now drowned, Old Reliable lives on as evidence that this part of Michigan once almost single-handedly fed the gluttonous copper demands of a youthful and growing America.

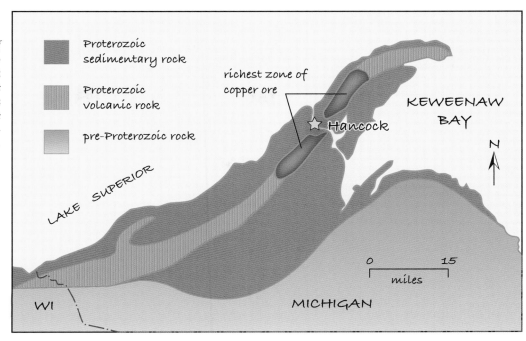

Geologic map of the Keweenaw Peninsula. The richest zone of copper ore is distributed across the heart of the historic mining community of Hancock.

This 10-pound chunk of pure amygdaloid copper is typical of the production material that allowed the Quincy Mine to deliver profits for fifty-three consecutive years. Six-inch ruler for scale. —Courtesy of Charles Herron, Wilmington, NC

The slant of the no. 2 shaft-rockhouse roof mimics the 54-degree angle of the Pewabic amygdaloid vein as it plunges into the subsurface.

36. Grand River Ledges, Michigan

42° 45' 35" North, 84° 45' 34" West
Pennsylvanian Period Sedimentation

A case of classic uniformitarianism read from rocks representing continuing environmental change.

Early geology texts are rife with controversy about how Earth has developed. Until the mid-eighteenth century, scholars championed catastrophism, the concept that Earth evolved through a succession of sudden, short-lived catastrophes—volcanic eruptions, earthquakes, and the effects of biblical flooding—each causing unprecedented biologic and topographic change.

In contrast, James Hutton, the father of geology, attributed Earth's profile to past forces that differed neither in kind nor energy from those operating in the present. His argument, known as the concept of uniformitarianism, is often abbreviated to "the present is the key to the past," meaning the study of the geologic processes presently shaping Earth's surface is the key to understanding how Earth has evolved over time. Neither of these doctrines has survived as it was originally conceptualized. The uniformitarian position dominates scholarly thinking today, but obviously the occasional catastrophic event, such as the 1980 eruption of Mount St. Helens, can and does occur.

Throughout the early and middle periods of the Paleozoic era, the slow development of an expanding structural dimple indented the crust of the Lower Peninsula of Michigan. By the time this widespread depression—the Michigan Basin—had matured, 15,000 feet of sedimentary rock, principally limestone and dolomite deposited within shallow seas, had filled it. On top of these marine deposits is an alternating sequence of 310-million-year-old strata deposited in floodplains, lagoons, swamps, and nearshore environments.

Outcrops of these Pennsylvanian-age rocks are uncommon in Michigan. Most are buried beneath the thin veneer of debris deposited when Quaternary-age glaciers retreated from the landscape some 12,000 years ago. An exception is the exposure of the Saginaw Formation—the largest outcrop of bedrock in central Michigan—along the banks of the Grand River and within nearby abandoned clay pits in northern Eaton County. Here, sediments were quietly and serenely deposited in fluctuating environmental conditions, leaving a heritage of strata that tell a classic tale of uniformitarianism.

The massive sandstone wall exposed along the Ledges Trail in Fitzgerald Park, on the northern edge of the town of Grand Ledge, is suggestive of an ancestral high-energy shoreline. The upper sections of the ledges are composed of thin crossbeds, which geologists interpret as evidence that wind-generated sand dunes migrated across a beach environment in response to seasonal weather patterns. At nearby Lincoln Brick Park, strata of the same age represent a drastically different environment. Thick layers of stream-deposited sandstone, units of gray limestone containing organic material, and even older deposits of low-grade coal containing terrestrial plant material indicate that this was once a poorly drained swamp surrounded by vegetated lowlands. In two nearby, privately owned quarry sites are shale beds containing fossils of the mud-loving brachiopod *Lingula* and siltstone with in situ plant roots, which together suggest a lagoon once separated the Fitzgerald beach and the swampy Lincoln Brick region.

These limited exposures are of great significance to scholars reconstructing the ecology of Eaton County during this one tick of geologic time. With succeeding ticks, conditions continued to fluctuate to and fro, each change accompanied by different rock types. This benign Grand Ledge story of depositional contrast and fluctuating environments is representative of thousands of other tranquil periods of time within the world of the uniformitarian practitioner. The occasional catastrophic event that interrupts such unobtrusive passages acts as scientific punctuation that separates one chapter of geologic history from another.

The ghostly remnants of crossbedding near the top of the sandstone wall fronting the Ledges Trail is evidence of a beach and nearshore environment.

Sandstone (top), gray limestone (bottom), and coal (not visible in this photograph) exposed at Lincoln Brick Park suggest these rocks were deposited in a sluggish river and fetid swamp environment. This site is located just north of Grand Ledge on Tallman Road. —Courtesy of Missy Norris, Eaton County Parks, MI

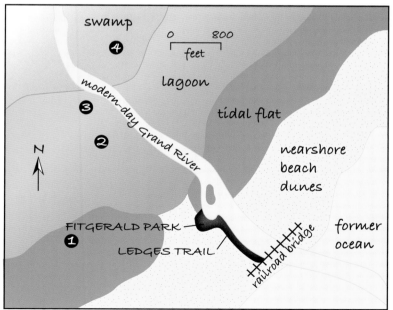

Map of the Grand Ledge region during the Pennsylvanian period showing the different environmental conditions. Fitzgerald Park (1) and Lincoln Brick Park (4) are open to the public. Sites 2 and 3 are quarries on private property. (Modified after Milstein, 1987.)

37. Sioux Quartzite, Minnesota

44° 00′ 48″ North, 96° 19′ 33″ West
Proterozoic Eon Metamorphism

Sandstone eroded from an ancestral mountain range hosts a clay seam of significance to Great Plains Indian tribes.

Alternating periods of crisis and quiescence punctuate the annals of geologic history just as they do human history. Events of a catastrophic nature are sandwiched between episodes during which Mother Nature seems to demand the right to catch her breath before being subjected to yet another onslaught of crisis. Consider the history of the Sioux quartzite, an unusual and very erosion-resistant rock that is erratically exposed throughout Minnesota, South Dakota, Iowa, and Nebraska. An excellent exposure of this famed formation sits at the center of Pipestone National Monument, 1 mile northwest of the community of Pipestone, in southwestern Minnesota.

A close examination of the composition and texture of the Sioux quartzite reveals the specific environment in which it was deposited, long before a mantle of plant and animal life had clothed the proto–North American continent. At Pipestone National Monument, this pink-hued quartzite envelops a thin layer of orangish red, metamorphosed claystone, prized for centuries by the Native American tribes that inhabit the northern Great Plains.

Approximately 1,850 million years ago, a range of imposing mountains extended in a generally east-west direction across today's Lake Superior region. The formation of these Penokean Mountains was but one phase of an ongoing, global episode of mountain building. Some 700 million years later, deep-seated extension forces racked the same region, giving birth to the Midcontinent Rift, an unsuccessful attempt to tear the youthful continent asunder. During the multimillion-year period of time bookended by these contrasting episodes of construction and destruction, when the Ponekeans were being reduced to mere hills and the forces of rifting had not yet gathered, the landscape lacked topographic appeal and the myriad processes of tectonic activity lay dormant. This was one of those interim periods when Mother Nature caught her breath, and the time when rivers deposited the Sioux quartzite.

The Sioux is a textbook example of quartzite, a metamorphosed sandstone composed almost exclusively of the mineral quartz that recrystallized during metamorphism. The quartzite's purity testifies to the sandstone having been subjected to prolonged and repeated periods of weathering, during which all chemically unstable minerals that were originally present in the rock decomposed due to interaction with atmospheric agents, leaving the very stable mineral quartz as the dominant material. The quartzite unit is distinguished by an interesting variety of sedimentary features, ranging from crossbeds and ripple marks to mud cracks, the latter confined to the few fine-grained beds contained within this 1,700-million-year-old formation. These depositional indicators strongly suggest that the sandstone that became the Sioux quartzite was deposited in braided river systems that sluggishly meandered across a low-relief landscape.

Since AD 1200, Native Americans have quarried the thin layers of metamorphosed claystone that give the Sioux quartzite ethnological importance. Named catlinite—also known as pipestone—in honor of George Catlin, an early artist of Missouri River Valley tribes, this 14-to-18-inch-thick, aluminum-rich layer of clay is one of the finest carving stones in the world. It is as soft as a fingernail and can be easily sculpted into pipes and other ceremonial objects. Its rusty color, attributed to the presence of hematite, an ore of iron, and its butter-smooth consistency are the primary reasons why it was a widely distributed tribal commodity. The confinement of this adaptable and malleable substance within layers of unyielding, brittle quartzite is hard-core proof that curious dissimilarities of rock often form the norm in nature.

Isolated exposures of Sioux quartzite are distributed across parts of four Midwestern states, surrounded by rock believed to be similar in composition and age.

Long before plant and animal life graced North America, rivers drained the land and deposited sediment, as indicated by the presence of crossbeds (dashed lines are examples) in the Sioux quartzite. The white blotches are lichen growth.

Close-up of a 15-inch-thick layer of catlinite enveloped by massive beds of Sioux quartzite on the grounds of Pipestone National Monument.

38. Thomson Dikes, Minnesota

46° 39′ 53″ North, 92° 24′ 16″ West
Proterozoic Eon Fissure Eruptions

An up close view of the plumbing system that built one of the thickest piles of lava known on Earth.

In comparison to other planets of the solar system, Earth appears to be unique. Thick thermal currents constantly churn its interior, plate tectonics movement continually re-arranges the geographic design of its crust, and oceans form and then are destroyed, while mountains and canyons are but ephemeral elements of the scenery. Translated into geologic headlines of the twenty-first century, this omnipresent interplay of land and water would read: Atlantic Ocean expanding, the Pacific shrinking; Alps and Himalayas growing, Appalachians declining; Africa splitting up, California slowly sliding offshore, and the Mediterranean being quietly squeezed shut.

An impressive single-frame imprint of this slow-motion geologic orchestration is visible from the Minnesota 210 bridge straddling the St. Louis River 1 mile east of Carlton. Multiple units of thinly bedded slates that comprise the Thomson Formation are exposed here as southerly plunging bedrock. These strata, originally deposited during the early Proterozoic eon as layers of mud and sand, were metamorphosed around 1,900 million years ago when they became involved in the Penokean orogeny, a mountain building event that stretched from Minnesota into Ontario. The mountain range itself has long since fallen prey to the inevitable forces of erosion and weathering, but one of its contorted roots is still visible as a small anticline in the slate beds immediately downriver of the bridge.

Following the birth of the Penokeans, this region seemed geologically quiet, but trouble was brewing. A deep-seated plume of molten rock under what is presently the central portion of Lake Superior exerted pressure upward, fracturing the Thomson Formation slate. Beginning 1,109 million years ago, these fissures functioned as conduits for the extrusion of magma onto the surface. Over the next 15 million years this hot-spot plumbing system poured forth an estimated 480,000 cubic miles of semifluid lava. When the flow of molten rock finally stopped and the magma remaining in the fissures had congealed as dikes—tabular igneous intrusions—a 10-mile-thick column of rock, comprised of hundreds of individual lava flows, had been extruded over an area equivalent to the size of Ohio. Recognized today as perhaps the world's largest igneous province by volume, it rivals the estimated volumes of the Deccan flows of India at 360,000 cubic miles, and the Siberian flows of Russia at 400,000 cubic miles.

An excellent exposure of this ancient plumbing system—a swarm of often high-relief, vertical dikes intersecting the Thomson Formation—is on view from the Minnesota 210 bridge. They range in thickness from several inches to 20 feet. The thickest one, alongside the cement wall to the right, viewed from the north side of the bridge looking upstream, displays horizontal columnar fractures that formed when the magma cooled. The cluster of dikes exposed at this geo-site is but one example of the many aggregations of dikes believed to be scattered throughout the region, most of them deeply buried. It is visible because the overlying basalt flows have been eroded away, exposing a network of filled geologic passageways, much like the standing pipes that remain after a house has been destroyed by a tornado.

For millions of years, periods of eruption created a barren landscape not unlike the surface of the moon. Today, this remaining dike-riddled tableau in Minnesota represents just one small note in the 4,600-million-year symphony of Earth history.

The small anticline in the Thomson Formation on the west bank of the St. Louis River immediately downstream from the Minnesota 210 bridge. The dashed line denotes the curve of this feature.
—Courtesy of Kristine Hiller, Jay Cooke State Park

This dike stands tall with 15-inch relief because these dikes weather more slowly than the adjacent metamorphic rock of the Thomson Formation. Car keys for scale.
—Courtesy of Kristine Hiller, Jay Cooke State Park

Close-up of a 15-inch-wide, 1,100-million-year-old dike (dashed lines mark its edges) cutting across the Thomson Formation. —Courtesy of Kristine Hiller, Jay Cooke State Park

39. Soudan Mine, Minnesota

47° 49′ 11″ North, 92° 14′ 31″ West
Archean Eon Mineralization

Descend 0.5 mile into the colossal, richest-of-all iron mine that nourished the mills of America for decades.

Anyone curious about the inner workings of mother Earth will find much of interest at Soudan Underground Mine State Park, 90 miles north of Duluth, in the Vermilion Iron Range. This is home to the Soudan, the oldest, deepest, and richest iron mine in the state. For eight decades it bustled with activity, feeding America's steel mills. Today, this 50-mile labyrinth is open to the public as a "cool," 52-degree-Fahrenheit opportunity to experience subterranean geology and the life of a hard-hat miner.

Tours begin at the eighty-year-old electric mine hoist of shaft no. 8, dropping 2,341 feet to the twenty-seventh level, the depth that was under development when the mine ceased commercial operation. This is not the route to the Earth's center, but the three long minutes visitors spend in the noisy cage that miners once used might give that impression. At the bottom, an electric train travels 0.75 mile to a cavernous room where the final ton of ore was extracted in 1962.

Science fiction writer Jules Verne used Mount Snaefell in Iceland as an entry point for his classic *Journey to the Center of the Earth*. There are similarities between his tale and that of the development of the Soudan Mine—the adventure of descending into the interior of Earth and the discovery of vast reserves of economic minerals. About 2,700 million years ago, volcanoes were active throughout northern Minnesota, littering what was then the ocean floor. Hot vapors and mineral-laden solutions periodically issued from the vents and made contact with the cold ocean water. The minerals in the solutions precipitated as a rock called banded iron formation. Banded iron formation is composed of alternating layers of hematite and magnetite and layers of red and white chert. The banded iron formation, however, did not become economically valuable ore rock until migrating hot waters later dissolved the chert, leaving a very dense, high-grade rock containing as much as 69 percent iron. The Soudan Mine is located in one of these pockets of high-grade rock.

The banded iron formation deposits were discovered during the Lake Vermilion Gold Rush of 1865. Unfortunately, the gold was merely fool's gold, but portions of the iron formation were seen to be high-grade in quality and massive in volume. Mining initially took place in vast open pits, but because the rocks were very dense and not conducive to fracturing, companies soon developed underground operations. As many as 1,800 men—Finns, Italians, Swedes, Austrians, Slovenians, and Cornish—removed a total of 15.5 million tons of ore before the Soudan closed. The ore is far from depleted, but competition from large-scale surface mines producing low-grade, iron-bearing rock called taconite, changes in steel production methodology, and the high costs of ore extraction forced the demise of underground operations.

The record depth of the Soudan Mine is the result of the orientation of the iron-bearing beds. Deposited originally at near-horizontal attitude, they were tilted to their near-vertical orientation during the Algoman orogeny, a mountain building event that occurred 2,700 million years ago. Erosion has obviously long since leveled those mountains.

The twenty-seventh level is also the location of the Soudan Underground Laboratory. Cutting-edge physics experiments are conducted here in a cosmic-radiation-free enviroment. The MINOS experiment examines neutrinos (massless subatomic particles) beamed underground from the Fermi National Accelerator Laboratory in Illinios, 450 miles to the south, while the CDMS experiment searches for evasive particles of dark matter, a substance supposedly formed moments after the Big Bang birth of the universe.

A fractured exposure of folded banded iron formation near the head frame of the Soudan Mine.
—Courtesy of Richard W. Ojakangas, Duluth, MN

Close-up of a 3-foot-long sample of high-grade iron, near the fence by the head frame, that is representative of the ore extracted from the Soudan Mine over a period of nearly eight decades. —Courtesy of Alyssa Wodtke, Oakland, CA

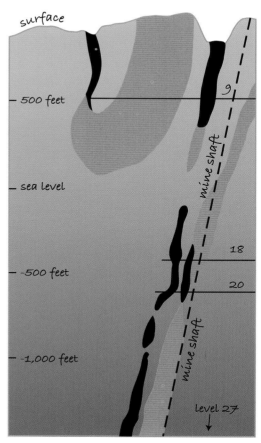

Cross section of the Soudan Mine, showing the relationship of ore (black), uneconomic iron-bearing rock (orange), and the greenstone country rock (green). Numbers identify depths of various mine levels. (Modified after Ojakangas, 2009.)

40. Petrified Forest, Mississippi

32° 31' 08" North, 90° 19' 23" West
Paleogene Period Fossilization

Experience a fossil hardwood and coniferous forest that shaded the countryside 36 million years ago.

Petrified wood commonly occurs with sedimentary rock. In one form or another petrified wood is the official gem, stone, or fossil of the states of Arizona, Louisiana, Mississippi, North Dakota, Texas, and Washington. Whether defined as a stone with intrinsic value, a small piece of rock, or evidence of prehistoric life, it is quite simply wood replaced by mineral matter.

For 90 percent of the history of the Earth, trees were absent from the landscape. Their roots did not anchor the soil, their branches did not shade the ground, and their massive bulk did not deflect the wind. The earliest fossil wood—palm tree–like trunks with fernlike branches—dates to the Middle Devonian period (see geo-site 60). Less than 100 million years later, trees had grown in variety and geographic extent to world dominance. Today, forests cover 30 percent of the land surface of Earth, evidence of a healthy planet.

There are two basic processes that explain the petrification of wood. Before either can occur, however, the tree must be preserved. Once a tree dies, quick burial is necessary so it is not subject to rot and decay. Even then, its ultimate fate may be alteration to coal. Identifiable plant remains characterize peat, an early stage of coal formation.

If, however, the fallen forest giant is covered by a concentration of the crystalline compound silica—for example, volcanic ash—petrification is practically assured. If silica-rich groundwater percolates through the burial site, the mineral-charged waters can either deposit silica in the voids separating the wood fibers, a process called permineralization, or silica can actually replace the tissue of the wood, an alteration termed replacement.

In both cases, silica is originally deposited as a gel. Over time, depending on the degree to which the groundwater is saturated with silica, dehydration may change the gel to a form of quartz: opal, chalcedony, jasper, or agate. When other elements are present, spectacular colors create dazzling varieties of petrified wood. Iron causes red, brown, and yellow hues; copper and chrome lead to green; aluminum makes white; and manganese dioxide colors the minerals black.

The best fossil wood site in the South, advertised as the third largest in the world, is about 2.5 miles southwest of Flora, off Petrified Forest Road. Here, in the Mississippi Petrified Forest, tree trunks continue to erode out of the badlands topography of the 36-million-year-old Forest Hill Formation, a sequence of sand and clay deposited in shallow water. Of the two types of wood that occur here, the better preserved is classified as coniferous—pine- and sprucelike—while the more abundant is a hardwood variety similar to maple and oak. A mixture of chalcedony and opal replaced the original organic material to fossilize these logs.

Originally, scientists believed these Paleogene-age trees grew in a northern climate before being leveled by a storm, swept into the area by regional river systems, and preserved as an ancient logjam—a thesis supported by the absence of limbs, bark, and root structures. Today, due to the tropical characteristics and general absence of growth rings in this fossil wood, consensus rallies around the portrait of a forest of 100-foot-high trees growing in a moist, warm climate and dying in place, having given shade, shelter, and food to a variety of vertebrate and invertebrate life-forms for a thousand years or more.

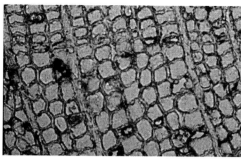

Close-up of the cell structure of a petrified coniferous tree, from a fossil log found at Mississippi Petrified Forest. Opal is the predominant mineral.
—Courtesy of the Mississippi Petrified Forest

The "Frog," a fossil remnant of a once stately tree, is reportedly lying in the precise spot in which it died 36 million years ago.

"Caveman's Bench" underwent a fair degree of weathering prior to its burial and preservation.

Magnified cell structure of a hardwood, from a fossil log found at Mississippi Petrified Forest. Chalcedony is the principal mineral. —Courtesy of the Mississippi Petrified Forest

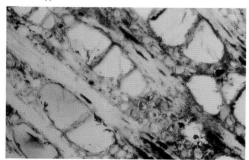

41. Elephant Rocks, Missouri

37° 39' 16" North, 90° 41' 18" West
Phanerozoic Eon Spheroidal Weathering

Is this a fossilized herd of ponderous pachyderms or an exotic demonstration of weathering?

Scientists attribute elephants' unusual memory to their having the largest brains, by mass, of all land mammals. Tipping the scales at a hefty adult weight of 10 pounds, this record brain size has conceivably enabled these masters of the Asian and African outback to record aspects of the past—for example, the location of a favored waterhole or a succulent fruit tree in season. America is not known for its elephants, but one herd decidedly deserves attention: the "pachyderms" of Elephant Rocks State Park, 4 miles south of Belleview off Missouri 21 in Iron County.

With a little imagination, the end-to-end row of giant boulders showcasing the park's topography appears as a trunk-to-tail assemblage of trained circus animals. Although they differ in composition from their flesh-and-blood namesakes, these wannabe elephants have also encoded past events. Since their memories are of geologic rather than biologic origin, the ingrained records involve not mere generations of time but the span of eras—in this specific instance, the Precambrian advent of the St. Francois Mountains.

Constructed of picturesque, red-hued rock that forms the geologic core of the Ozark Mountains and the oldest Midwestern landscape, the St. Francois Mountains dominate the scenic beauty of southeastern Missouri. This hard-rock terrain is the sole remnant of a mountain range birthed by continental plate collision during Precambrian time. As these mountains muttered and heaved to life—growing in stature and prominence—friction-generated heat gradually replaced their solid subterranean roots with expanding chambers of molten rock.

The subsurface melt crystallized into granite 1,485 million years ago, a birth date encoded radioactively into its chemical fabric. Granite's light color and mineral content—composed of common quartz and potassium- and sodium-bearing feldspar—are a reflection of the acidic nature of its parental magma. As the magma cooled and contracted, intersecting sets of vertical fractures dissected the developing granite. Then, layers of overlying sedimentary rock were gradually eroded away, and the myriad forces of weathering attacked the coarse crystal hide of the exhumed granite, increasing both the width and depth of its many fractures.

The curious, splintered maze on view in Elephant Rocks State Park is a classic example of a tor—an isolated pinnacle of much-jointed granitic rock that, through exposure to intense weathering, assumes eccentric or fantastic shapes. The corners and edges of the exposed granite blocks decompose faster than the flat surfaces, becoming rounded and geometrically muted in a process known as spheroidal weathering. Over time the number of elephants inhabiting the park will change as old ones weather to sand and clay and newborns develop from the fractured hillside.

Commercially known as "Missouri red," the granite in the Elephant Rocks area has been quarried since 1869. Perhaps its most historic use is as facing stone for the piers of the Eads Bridge, an iconic structure in the city of St. Louis. Ironically, when this arch bridge—the longest of its kind at the time—was opened across the Mississippi River in 1874, a circus elephant was led across the 6,442-foot-long span to test its soundness, in part because it was believed that elephants had instincts that would keep them from setting foot on unsafe structures.

"Dumbo," the 680-ton, 35-by-27-foot matriarch of the elephant herd, is a beautiful example of the weathering that has deeply subdivided this exhumed mass of granite.

Granite exposures in Elephant Rocks State Park (star) and the St. Francois Mountains form an off-center bull's-eye within the Ozark Mountain region.

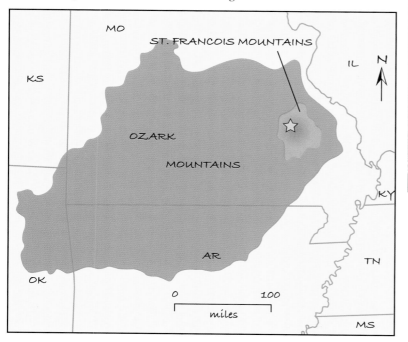

Only a trace of the original angularity of these granite boulders exists, although the planes of the fractures that separated the boulders are still evident.

42. Grassy Mountain Nonconformity, Missouri

37° 34' 01" North, 90° 20' 55" West
Proterozoic Eon–Paleozoic Era Nonconformity Development

Within this 980-million-year-old gap of forever-lost time, Earth experienced unprecedented changes.

At the rate new geologic information is presently being gathered, the day is approaching when *The History of Earth* might need to be published as a two-volume set. A logical dividing line between these texts would be the transition from Precambrian time to Phanerozoic time 542 million years ago. This milepost of time marks unprecedented change in atmospheric composition, meteorological conditions, and the nature and distribution of life. Within the United States, geologic evidence of this transition is confined principally to rugged terrain in the Rockies and Appalachian Mountains.

An exception exists in Madison County near Junction City, along both sides of Missouri 72, 1.7 miles west of its intersection with US 67. In this unusual roadcut, in which sedimentary strata belonging to the 500-million-year-old Bonneterre Formation overlie the eroded surface of the 1,480-million-year-old Grassy Mountain ignimbrite, a nearly vertical dike highlights the rock contrast. A layer of boulders several feet thick divides the two suites of rock.

The story of this roadcut can be reduced to five stages: (1) An explosive volcanic eruption produced an ash flow that cooled and crystallized 1,480 million years ago into a fractured black rock known as ignimbrite. (2) Magma intruded the fractures of the Grassy Mountain ignimbrite 150 million years later. Eventually the molten slurry cooled, forming the crosscutting, 5-foot-wide dike. (3) Both the ignimbrite and dike experienced an extensive period of weathering, but the exposure south of the road illustrates how much more the intruded rock was affected. The dike weathered to a deeper degree than the ignimbrite because its basic chemical nature is more susceptible to weathering than the acidic nature of the ignimbrite. A coarse mantle of angular rock debris formed as the ignimbrite and the dike weathered. (4) A shallow sea invaded southeastern Missouri from the south some 500 million years ago. Waves and currents rounded the angular debris, forming a rock called conglomerate. (5) Lastly, the layered Bonneterre Formation was deposited across the seafloor as calcium carbonate precipitated from the seawater. This rock unit hosts lead, copper, and zinc that once supported a viable regional mining industry. The minerals precipitated from hot waters circulating upward from deep-seated magma activity.

The key to this interpretation, and to the understanding of this geologic site, is the period of weathering (stage 3) that occurred, which is significant because of the rock that is missing. This contact between the conglomerate and ignimbrite is a stellar example of a nonconformity—a break in the geologic record formed by the juxtaposition of younger sedimentary rock with older, weathered igneous rock. During this 980-million-year hiatus—1,480 million minus 500 million years—hard-shelled marine animals evolved from those with soft tissues; at least two glacial episodes ensued, one supposedly reaching the equator and creating Snowball Earth (see Proterozoic Eon in the introduction); and a dramatic change in the chemistry of the atmosphere became the driving force for the Cambrian explosion, a period of unprecedented diversification of life.

You can easily touch the Grassy Mountain nonconformity, but the human mind finds it difficult to grasp its contained loss. Here, the geologic record is forever incomplete, a victim of the relentless forces of weathering and erosion. This black hole of information, this missing 980-million-year page of geologic history, can be filled only with supposition, never with fact.

View of the contact between ignimbrite bedrock and the fractured dike. Here, on the north side of Missouri 72, the conglomerate and the Bonneterre Formation have eroded away.

Exposure of the Grassy Mountain nonconformity (dashed line) in the Missouri 72 roadcut, on the south side of the road.

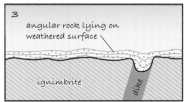
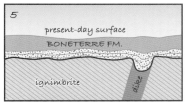
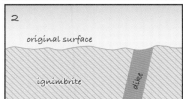

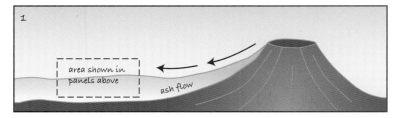

Step-by-step chronicle of the Missouri 72 roadcut. Note how the conglomerate sagged into the dike, a stark indication that the dike weathered faster than the ignimbrite. The heavy black line is the plane of the nonconformity.

43. Chief Mountain, Montana

48° 55′ 57″ North, 113° 36′ 35″ West
Mesozoic Era Mountain Building

This exception to the principle of superposition is explained by the birth of the Rocky Mountains.

The high prairie of northern Montana rolls westward in an unfolding panorama of speckled sage and open sky, defined by a slowly rising but otherwise benign landscape. Along Montana 17 a few miles southeast of the Canadian border, topographic change is finally realized by the haughty appearance of Chief Mountain, a sentinel that identifies the Rocky Mountain Front, the eastern border of Glacier National Park, and the location of one of the world's classic—albeit enigmatic—geologic structures.

A model klippe—an outcrop isolated by faulting and erosion—standing 9,006 feet high with 1,500 feet of relief, Chief Mountain is constructed of sedimentary rocks that are of a different age and environment than the rocks of the underlying terrain. A well drilled from the mountain's crest would first engage 1,300-million-year-old green shale of the Appekunny Formation and then 1,450-million-year-old tan and red strata of the Altyn limestone. These two formations make up a significant portion of the Belt Supergroup, perhaps the best-preserved sequence of middle Proterozoic–age rock in the world. Close examination of these layers uncovers beautifully preserved ripple marks, mud cracks, raindrop impressions, and fossils of several species of algae. The fossils and preserved sedimentary features are evidence that the strata were deposited in both playa and perennial lake-basin environments that periodically were connected to the open ocean. Finally, the drill bit would penetrate sandstone and shale beds deposited under marine conditions a mere 100 million years ago, during the Cretaceous period.

The enigma of Chief Mountain has nothing to do with either the different depositional conditions or the 1,350-million-year age difference between these two suites of rocks. Instead, it is the *age relationship* of these juxtaposed formations. In Chief Mountain, ancient rocks overlie much younger strata, contrary to the logic of a basic geologic axiom. As mentioned in the introduction, according to the principle of superposition, in an undisturbed sequence of strata the oldest beds are positioned on the bottom. How is this deviation from the venerated principle explained?

Answers became available when geologists recognized that the northern Rocky Mountain terrain is composed of thick sequences of faulted and stacked sedimentary rocks. The nature of the faults suggests that as the forces compressing the crust in this region increased in intensity, the faults became overthrusted upon each other, much like the arrangement of cedar shakes covering a roof. This fracturing and subsequent stacking of strata effectively reversed the age relationship of the rocks—older rocks were placed over younger ones.

The rocks of Chief Mountain were thrust up and over younger rocks sometime between 170 million years ago, during the construction of the ancestral Rocky Mountains, and 70 million years ago, when the modern Rocky Mountains were forming. During this 100-million-year time frame the Lewis Overthrust slab formed. Measuring 200 miles long and more than 2 miles thick, this massive slab of rock was transported eastward an amazing 50 miles along one of the largest thrust faults in the world. Since then, erosion has altered the slab so that today Chief Mountain stands completely isolated—but no longer an enigma.

Because the chapter and verse of its geologic history are so beautifully preserved, the Chief Mountain klippe is widely recognized as an ancient and revealing window through which the process of modern mountain building, such as that underway today in the Himalaya Range of southern Asia, can be better understood.

Though it has gone by many names, including Kings Peak and Tower Mountain, Chief Mountain was derived from the Blackfeet Indians' name for this peak.

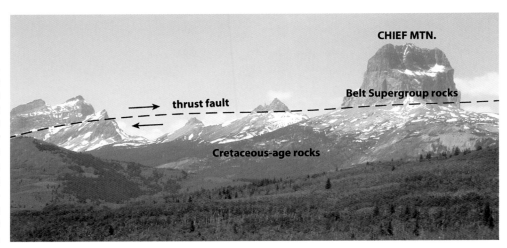

At Chief Mountain, limestone and shale formations of the Proterozoic-age Belt Supergroup overly much younger Cretaceous-age rocks, a textbook example of a thrust fault. Arrows indicate relative rock movement along the fault plane.

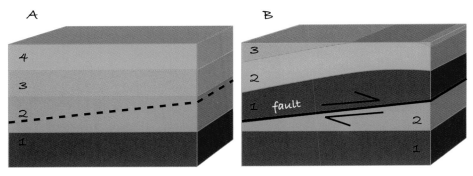

Before faulting (A), rock formations lay in a normal sequence, youngest (4) overlying oldest (1). Thrust faulting (B) reversed their positions, placing older strata above younger strata. Arrows indicate relative rock movement.

44. Madison Slide, Montana

44° 49' 50" North, 111° 25' 32" West
Holocene Epoch Mass Movement

This mountain-in-motion site bears witness to gravity's awesome power.

Near midnight on August 17, 1959, a 130-acre slab of 2,700-million-year-old weathered schist and dolomite abruptly broke loose from the south wall of the Madison River canyon in Montana. Within seconds, 28 million cubic yards of fractured rock had crossed the canyon—at speeds exceeding 90 miles per hour—and raced 400 feet up the opposite slope. Once the dust had settled, a portion of the canyon was filled with rubble to depths of 150 to 350 feet, twenty-eight people had died, and the debris-damned Madison River was forming a new lake.

This event bears chaotic testimony to the fact that of the planets bound by gravity to the sun, Earth appears to be unique. It is geologically alive. Within its core and mantle, temperature and pressure surges of intense magnitude continually create chemical and physical change, breaking Earth's relatively thin crust into major and minor tectonic plates that for the better part of 4 billion years have been drifting and shifting in a three-dimensional, musical-chair fashion. The effect of all this tectonic turmoil is that Earth's surface continually falls victim to two elemental forces that seek individual dominance: construction in the form of mountain building, and destruction in the form of erosion and weathering.

The principle forces of erosion are those of wind, ice, and water, which do their work in active, flowing environments. Given enough time, these three can and will reduce the highest-elevation rock to a minute grain of sand. Often overlooked is the passive force of gravity, the bane of the aging human body *and* Mother Nature. When gravity is in the driver's seat of erosion and is accompanied by water, mass movement becomes the name of the game.

The balance between factors that encourage movement and those that resist it determines the probability of mass movement. Steep slope angles and the lubricating effect of water create situations of instability. In contrast, low angles and dense vegetation discourage movement. In the Madison River canyon, an apparent mixed-bag relationship has long existed between these conditions. One factor, however, tipped the scale in favor of gravity-fueled slope failure: the layering of the bedrock along the south side of the canyon parallels the angle of the slope, providing surfaces for rock to slide into the canyon. The slope was primed for a jarring action to trigger a disaster.

The trigger was pulled at exactly 11:37 on that fateful night. North of West Yellowstone, Montana, and at a depth of 9 miles, the Precambrian heart of the Madison Range ruptured along a fault. Seismic waves capable of destroying the best-built structures radiated outward from this epicenter in a three-dimensional pattern. The rupture, the largest earthquake ever recorded in the northern Rocky Mountain region, with a magnitude of 7.5 on the Richter scale, became catastrophic history.

In less than sixty seconds the landscape changed radically. About 5 miles east of the slide, the floor of Hebgen Lake warped 9 feet, and for twelve hours following the earthquake huge waves sloshed back and forth across the lake surface. Ground ruptures rippled through the Red Canyon and Cabin Creek areas northeast of the slide, displacing their elevations in places, respectively, by a maximum of 22 and 10 feet. Downstream, 80 million tons of broken rock avalanched across the canyon of the Madison River.

The Earthquake Lake Visitor Center is located on slide debris, close to the very center of this topographic memorial to the destructive power of gravity. Here, a series of educational displays preserve chilling details of the largest earthquake-triggered landslide in recorded North American history.

The slide denuded the canyon wall (left), supplying a colossal mass of rock and forest debris that dammed the Madison River and created Earthquake Lake.
—Courtesy of the U. S. Geological Survey

Before the earthquake, the layers of Memorial Boulder paralleled the slope of the Madison River canyon's north wall. Layering such as this in slopes, along with the lubricating effect of water, has long been a formula for mass movement disaster.

The 10-foot-high trace of the Hebgen Fault slices through the Cabin Creek area, 5 miles northeast of the Madison slide. The ground ruptured here, creating this scarp, during the 1959 earthquake.

45. Butte Pluton, Montana

46° 01′ 49″ North, 112° 30′ 37″ West
Cretaceous Period Mineralization

Haloes of copper, zinc, and silver once radiated outward from this richest hill on Earth.

Seventy-five million years ago an igneous intrusion penetrated western Montana. Such was the scope of its influence that it strengthened the backbone of the then-developing modern Rocky Mountains to the status of near invincibility. Once emplaced and crystallized, the base of this former cauldron of molten crust and mantle lay at a depth of approximately 10 miles and measured some 75 by 25 miles. Called the Boulder Batholith, this mass of acidic rock occupies a volume of nearly 19,000 cubic miles, more than three times the amount of water in Russia's Lake Baikal, the world's largest freshwater lake by volume. Erosion has exposed the batholith at the surface today.

The area of the Boulder Batholith was destined to become one of the most economically valuable mining districts of the world. It is constructed of fifteen major plutons, subunits of rock characterized by slightly different ages, textures, and chemistries. The most important and voluminous of these is the Butte Pluton. This granite mass was endowed with minerals that, for many years, were exceeded in value only by gold deposits extracted from the East Rand Mine in South Africa. Its economic heart lies beneath the many streets and buildings of the city of Butte.

Butte's economic notoriety began in 1864 with the discovery of gold nuggets in local stream gravels. Soon the focus shifted to the plethora of veins that cut through the granite, each filled with minerals rich in silver, molybdenum, manganese, zinc, and lead all locked in a gangue (economically worthless rock and mineral) of quartz and fool's gold. Great concentrations of copper, an element considered worthless in the early days of mining, bound the ore (economically valuable rock and mineral) together. By the 1890s, however, due to the electrification of America afoot, more than three hundred underground mines in the Butte Pluton were producing 40 percent of the world's supply of copper.

The history of this mining district is divided into two phases. The first lasted more than a century, during which roughly 10,000 miles of shafts and tunnels—dug first by pick and shovel and then drill and dynamite—snaked throughout a three-dimensional hodgepodge of sets of veins. The mineral veins formed in fractures in the pluton that developed as it cooled and contracted. Minerals precipitated from hot water—sourced from the bowels of the Earth—as it circulated around the hot core of the pluton. Two sets—the Anaconda and the Blue—are constructed of veins measuring up to 100 feet wide and more than 1 mile in length and depth. Together they have contributed 80 percent of the area's vein-ore production. The mineralization that occurred in the vein sets follows a predictable geologic pattern: copper near the pluton core, zinc and lead at an intermediate distance, and manganese and silver within the peripheral zone.

The second phase of mining, initiated in 1955, ushered in the era of open-pit operations, which extracted tremendous volumes of low-grade ore concentrated over the core of the pluton. Over a quarter century, mine operations processed 1.4 billion tons of rock averaging less than 1 percent metal content, the principle elements being copper and molybdenum. Operations closed in 1982, the victim of low copper prices and increasing use of aluminum in the electrical industry. Today, the public can view the footprint of the open-pit operations: the massive Berkeley Pit is about 1,800 feet deep and averages 1 mile in diameter.

The Butte Pluton has contributed 10 million tons of copper to the growth of America, plus enormous amounts of other categories of metallic minerals. Sporadic mining operations continue, evidence that this massive granite intrusion is still capable of adding to its already heady reputation as "the richest hill on Earth."

One of a series of sheetlike intrusions that constitute the Anaconda set of metal-bearing veins. This exposure is located at the intersection of Jackson and Gold streets in Butte.

Thirty-five billion gallons of water with the acidity of vinegar fill the Berkeley Pit, a highly visible reminder of the era of surface mining. The water level is still rising. —Courtesy of the Butte-Silverbow Chamber of Commerce

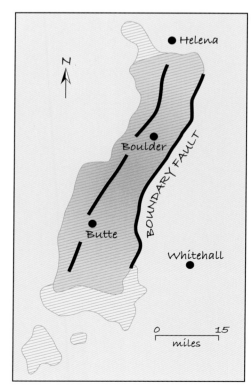

Granite of the Butte Pluton (pink) comprises the majority of the exposure of the Boulder Batholith (hachured). Fault systems (black lines) cross this massive structure.

109

46. Quad Creek Quartzite, Montana

45° 05′ 32″ North, 109° 19′ 12″ West
Archean Eon Continental Birth

This quartzite holds the key to understanding the origin of the North American continent.

While the presentation of significant events listed in chronological order—the first fish, first tree, or first mammal—is the conventional way to tell the story of the 4,600-million-year history of Earth, other revealing approaches are available. One might emphasize the serial construction and destruction of individual mountain ranges over time: orogeny versus erosion. Another could highlight long-term shifts between warm and cool climates: desertification versus glaciation. Chronologies of this nature historically have reflected a continental scale.

With the advent of the plate tectonics theory, geologic histories have been commonly woven to reflect a much larger scale. These stories emphasize the creation and destruction of ocean basins as globally distributed landmasses periodically come together and then break apart—the aptly named supercontinent cycle. The supercontinent cycle theory suggests that all the landmasses of Earth periodically assemble as one and then fragment in a type of universal maneuver much like a western hoedown, in which dancers gather into a circle and then retreat. While the scientific jury is divided on just how many supercontinents have existed over time, at least four—along with several predecessors—are generally accepted. From youngest to oldest, they are:

- Pangaea: assembled 250 million years ago and now in a mature stage of dissection.
- Rodinia: formed around 1,000 million years ago by the joining of early versions of South America, eastern North America, and western Africa.
- Nena: assembled 1,800 million years ago, its name an acronym for the joining of ancestral Northern Europe and North America.
- Arctica: the union 2,700 million years ago of the three ancestral continents of Siberia, Greenland, and Kenorland—the latter commonly referenced as a very early version of proto–North America.

Our understanding of the succession of supercontinents can be traced to the work of hundreds of researchers who have radiometrically age dated thousands of rock samples.

Naturally, more is known of the youthful Pangaea than the elderly Arctica. But even Arctica's history is discernible. One limb of its family tree is traced back to the amalgamation of four ancient terranes—Superior, Rae/Hearne, Slave, and Wyoming—into Kenorland, the rock core of present-day North America. Three of these are of special interest, because their boundaries envelop exposures of the oldest rocks discovered to date in both Canada and the United States. The analysis of these so-called genesis rocks is the key to understanding the very earliest pages of geologic history.

The Nuvvuagittuq greenstone, the earliest representative of the Superior Province of Canada, lays claims to the title of oldest Earth rock. Its 4,280-to-3,800-million-year age literally spans the time division between the Hadean and Archean eons. Many geologists bestow runner-up status to the 4,000-to-3,940-milllion-year-old Acasta gneiss from the Slave Province. The overlap in the range of ages of these two contenders, however, creates doubt as to which is really the most elderly. Little debate exists, however, when the subject shifts to the oldest U.S. rock: the Quad Creek quartzite. Exposed across the Beartooth Plateau, which straddles the Montana-Wyoming border, this rock, which is no older than 3,960 million years, forms the nucleus of the Wyoming Province.

Unlike the Canadian exposures, Quad Creek samples are roadside accessible, specifically 200 feet south of mile marker 62, along the west side of US 212 south of Red Lodge. Advertised as "the most beautiful drive in America," this stretch of US 212, also called the Beartooth Highway, should also be lauded as the "gateway to the Quad Creek quartzite." Through the study of this Methuselah-like ancestor and its darker-toned Canadian cousins, geoscientists have identified the ancient Wyoming, Superior, and Slave microcontinents that, with the Rae/Hearne Province, form the core of present-day North America. In turn, these microcontinents have aided geoscientists in their understanding of supercontinent cycles and how supercontinents—Arctica, Nena, Rodinia, and Pangaea—evolved over time and finally, upon the breakup of Pangaea, into present-day North America and the globe's other continents.

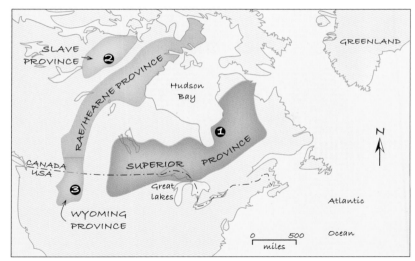

Geologists have found "senior citizen" rocks that identify three of the four ancient terranes that came together to first form Kenorland and then through collision with Siberia and Greenland to form the early supercontinent of Arctica: (1) Nuvvuagittuq greenstone, (2) Acasta gneiss, and (3) Quad Creek quartzite. The Rae/Hearne Province has been identified by rock samples, but none are of a record age.

Two of the oldest antiques known on Earth: the Acasta gneiss (left) and Quad Creek quartzite. Six-inch ruler for scale. —Courtesy of Charles Herron, Wilmington, NC

A dark intrusion cuts the center of the Quad Creek quartzite, which represents some of the very first solid planetary material that formed after the molten Earth began to cool.

47. Ashfall Fossil Beds, Nebraska

42° 25' 13" North, 98° 09' 21" West
Miocene Epoch Fossilization

Mass death and quick preservation at a favorite water hole.

Around the northeastern Nebraska community of Plainview, near the western limit of soils deposited by the ebb and flow of Pleistocene ice sheets, the land rises from the gentle slopes of what's called the Dissected Till Plains to the more diverse and hilly terrain of the Great Plains. This is tried-and-true Cornhusker country. It is also the land of the ubiquitous cow and coyote, the occasional bison, the melodic meadowlark, and the iconic bald eagle.

Twenty miles northwest of Plainview, the landscape remains the same, but strangely the animal population does not. Here, a common listing of resident life-forms would include herds of rhinoceroses and elephants, packs of wild dogs, flights of gregarious crowned cranes, and the pedestrian secretary bird. In both imagery and description, this population seems more like the veldt of southern Africa than the prairie of Nebraska. And yet the eye does not deceive. Ample evidence lies within exposures of the Ash Hollow Formation of middle Miocene age. The Ashfall Fossil Beds State Historical Park houses one of the richest paleontological finds ever discovered, with exquisitely preserved specimens of vertebrae, teeth, tusks, and ribs, along with tracks.

Since its discovery on the edge of a cornfield in 1971, scientists have uncovered the remains of at least thirteen different mammals. There are numerous tooth-scarred, disarticulated bones, the result of predatory scavengers having ravaged and scattered the carcasses. Many of the more than 350 skeletons that have been exhumed are in three-dimensional, in situ posture. Llamalike camels, three-toed and single-hoofed horses, horned deer, four-tusk elephants, and saber-tooted cat bones exist in scenes of mass congestion. The remains of smaller animals, such as tortoises, birds, moles, and an articulated snake, complete the catalogue of death.

One of the more dramatic revelations involves *Teleoceras*, a hippopotamus-like, herbivorous, stubby-legged, barrel-bodied type of rhinoceros that supposedly called both land and water home. These fossils occur in herdlike groups, many in moment-of-death poses. The five-to-one ratio of female to male skeletons suggests they lived in haremlike association. Two discoveries are perhaps unprecedented in the annals of paleontology: a pregnant female skeleton encasing the bones of her unborn offspring and a huddled mother and infant in a position suggestive of nursing.

This remarkable treasure trove of fossils is preserved in a layer of fine, gray volcanic ash composed of microscopic shards of broken glass. Chemical fingerprinting of the ash suggests it came from an extinct Idaho volcano that erupted 11.83 million years ago. Studies indicate this eruption was one hundred times more powerful than the 1980 explosion of Mount St. Helens.

Immediately after the eruption, the resulting ominous cloud of abrasive material began its easterly jet-stream journey, blanketing the countryside. In Nebraska, myriads of animals were gathered about a favorite water hole, quenching their thirst, all the while ingesting the powdery material. The smaller animals died first, the larger later, by asphyxiation as the shards of glass tore into their lungs and destroyed their respiratory systems. The rhinoceroses were the last to expire, their remains crowded around the outer edge of the water hole. Ultimately, the same ash that created this scene of mass death entombed it, forming a Pompeii-like time capsule of prehistoric life that is on display from May to mid-October in the Hubbard Rhino Barn.

"Sandy" (no. 3) has lain next to her month-old calf ever since her death almost 12 million years ago.

The ash beds of this "prairie Pompeii" site have to date produced the remains of more than one hundred fossilized rhinoceroses.

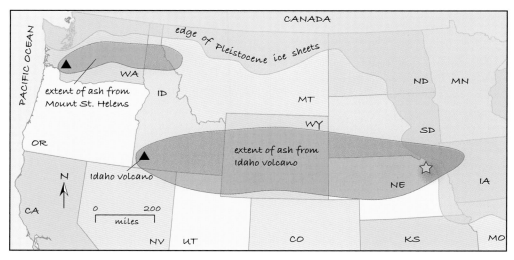

Ash from the Idaho eruption 11.83 million years ago spewed over seven states, across an area several times larger than that affected by the 1980 eruption of Mount St. Helens. The ash beds at Ashfall Fossil Beds State Historical Park (star) exist today because they lay beyond the edge of the destructive ice sheets of the Pleistocene epoch. (Modified after Rose et al., 2003.)

113

48. Scotts Bluff, Nebraska

41° 50' 08" North, 103° 42' 03" West
Cenozoic Era Terrestrial Deposition

A High Plains chain of topographic mile-markers tells a tale of terrestrial deposition and erosion.

To the intrepid nineteenth-century travelers who chose to "go west, young man," the passage across the grassland-cloaked sand hills of central Nebraska surely appeared never ending. The travelers gauged their 3-mile-per-hour progress by using pioneer-era mile markers that lay upon the ever-distant horizon: Courthouse and Jail Rocks, then Chimney Rock, and finally Scotts Bluff. Lying within sight of Nebraska 92, southwest of its namesake city, Scotts Bluff—"hill hard to get around" to the Plains Indians—is now a national monument, in large part because its north face contains more geologic history than any other site in Nebraska.

Deposited during the Oligocene epoch of the Paleogene period, the strata forming Scotts Bluff measure 800 vertical feet from the banks of the North Platte River to the 4,659-foot-high tourist overlook. Based on age span and thickness, simple mathematics indicate deposition of these sediments occurred at a rate of less than 0.1 inch over the life span of the average person. In sharp contrast, they have eroded some 12 inches in the past seventy-five years. The uncomplicated layer-cake, sedimentary-rock construction of Scotts Bluff indicates no tectonic folding, faulting, or metamorphism has disturbed these rocks. They are a textbook example of a basic axiom of geology, the principle of original horizontality: sediments are deposited in layers parallel to the Earth's surface, making them as easy to read and study as an open book lying flat on a table.

Three formations comprise Scotts Bluff, and each one contains evidence of the pervasive semiarid, nonmarine environment that dominated western Nebraska 33 to 22 million years ago. At the base, the tiered, grass-covered slope is composed of pink and grayish green siltstone of the Brule Formation. These massive beds envelop copious volumes of pyroclastic debris deposited by dense clouds of volcanic ash that drifted into the area from sources within the then still-rising Rocky Mountains. A variety of animal fossils abound, but plant remains are relatively rare. Overall, these findings suggest that the Brule was deposited in a prairielike floodplain landscape carved by rivers and colonized by a rich blend of mammals, freshwater reptiles, and even the occasional bird.

The brown and gray exposures of the cliffs of Scotts Bluff characterize the Gering and overlying Monroe Creek–Harrison formations. These massive sandstone units display extensively developed crossbedded structures, thought to be the product of both wind-driven sand dunes and ephemeral braided river systems. The occasional burrow-riddled bed suggests beetlelike insects had infiltrated the area.

The environments that gave birth to the strata found here can be extended 35 miles east to Chimney Rock and Courthouse and Jail Rock by another axiom of geology: the principle of lateral continuity. This principle posits that strata of similar composition separated by an erosional feature, such as a river valley, are assumed to have been originally continuous. This collection of outliers represents the vestiges of an ancestral Great Plains that relentless wind, rain, and gravity have incised over the past 22 million years. In due process the same forces of erosion will completely strip away these isolated rock columns; indeed, the spire of Chimney Rock has reportedly lost 30 feet in 150 years. Some day in the future, these few remaining pages will be forever torn from the book of time.

Either wind or flowing water, or both, deposited the crossbedded sandstone found adjacent to the concrete steps along the 1.6-mile-long Saddle Rock Trail.

The uncomplicated layer-cake geology of Scotts Bluff is evidence that tectonic forces have not yet altered some portions of North America.

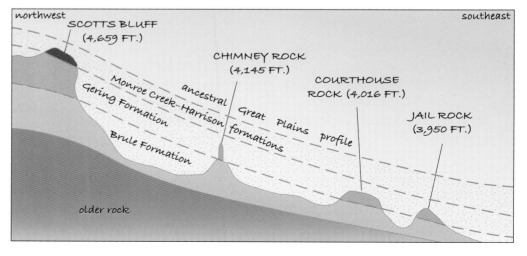

Cross section showing how the principle of lateral continuity is employed to reconstruct the original distribution (dashed lines) of the three formations seen at Scotts Bluff.

49. Crow Creek Marlstone, Nebraska

42° 46′ 18″ North, 98° 04′ 03″ West
Cretaceous Period Tsunami Deposition

An unassuming lens of rock is evidence of a catastrophic meteorite impact and tsunami waves.

Unfettered by forest and mountain, the featureless landscape of Nebraska is cleansed by currents of air so pervasive that it has the sixth largest wind energy potential of any state. This unsheltered, gusty scenario is in contrast to that prevailing some 74 million years ago, when the region was about to be inundated by a tsunami of celestial derivation. Toward the dusk of the Cretaceous period, the Western Interior Seaway, a 600-mile-wide sea extending from the Arctic Ocean to the Gulf of Mexico, drowned Nebraska and much of the rest of North America's midregion. Of the formations deposited in the fetid depths of this seaway, one of the most recognizable is the Pierre shale, a fine-grained sedimentary rock. It has two distinctive characteristics: an all-encompassing black color—attributed to a high organic content—and a tendency to slump and slide when wet.

Normally Pierre shale underlies the state at considerable depths. Exposures of a provocative nature, however, occur along the banks of the Missouri River in Niobrara State Park, immediately west of Niobrara. Here, a 3-to-10-foot-thick, cream-colored bed composed of marl (calcium carbonate mixed with clay), gravel-sized quartz, sandy siltstone (rock with a larger percentage of silt than sand), and a layer of shale fragments smudges the signature hue of the Pierre with a texture akin to that of a marbled Bundt cake. Called the Crow Creek marlstone, this rock's features indicate it was deposited in a high-energy water environment. This particular juxtaposition of strata, one deposited in quiet deep water (Pierre) and the other in disturbed shallow water (Crow Creek), is a geologic enigma that begs for solution.

After considerable study, geologists determined that the gravel-sized quartz grains in the marlstone contain evidence of celestial collision: their crystallographic framework was altered to include parallel deformation features, anomalies caused by compression forces associated only with meteorite impact. The veil of confusion lifted, and the search for the proverbial smoking gun to explain the origin of the marlstone began. What happened?

Around 74 million years ago the Cretaceous skies over north-central Iowa glowed in incandescent brilliance, lit by a 2-mile-wide, 10-billion-ton, stony meteorite that flashed 45,000 miles per hour through the atmosphere. When it collided with Earth near the eastern shoreline of the Western Interior Seaway, it penetrated the bedrock to a depth of 1 mile, creating a crater 25 miles wide. Instantaneously, a massive fireball radiated outward, incinerating trees for 300 miles and carving a radius of death for 650 miles.

Simultaneously, the surface of the seaway roiled upward into shock-induced tsunami waves that moved through the water at the speed of sound. Within minutes they had entered northeastern Nebraska and approached the south slope of Sioux Ridge, an ancient submarine ridge. As the tsunami waves moved into shallow water, their bases bulldozed and scoured the seafloor, ripping up chunks of bedrock—suspended debris that became the "stuff" of the Crow Creek marlstone. This particular and unusual marlstone is a rare example of a rock born of tsunami turmoil. The actual meteorite crater, called the Manson, lies 200 miles east of Niobrara State Park, buried beneath glacial drift in Iowa. Radiometric dating revealed the crater and marlstone share a common age of 74 million years, putting to rest any remaining doubt regarding their intimate relationship.

Could this scenario of cause and effect occur again? Certainly! In 2009, the asteroid DD45 buzzed the Earth at a distance of roughly 48,800 miles—only double the height of some telecommunications satellites. Another exploded over Siberia in 1908, leveling 800 square miles of forest. "When?" is the only question regarding the next collision of Earth and meteorite.

Cream-colored exposure of Crow Creek marlstone on the Missouri River trail, 2.25 miles west of the Niobrara State Park entrance along Nebraska 12.
—Courtesy of Rich Lofthus, Yankton, SD

Thin-section comparison of (left) quartz-bearing rock in the natural state (white crystal) and a similar rock altered by extraterrestrial impact (striated crystal). —Courtesy of Ray Anderson, Iowa Geological Survey

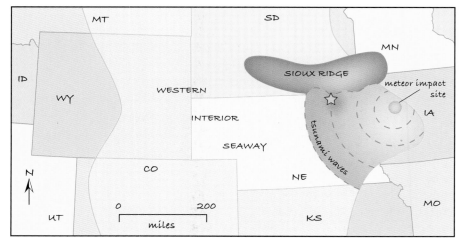

Dashed lines demonstrate the passage of tsunami waves as they raced from the meteor impact site across the Western Interior Seaway and then crashed ashore on Sioux Ridge at Niobrara State Park (star).

50. Sand Mountain, Nevada

39° 18' 31" North, 118° 23' 49" West
Quaternary Period Dune Development

Come early for a front row seat for the next performance of Symphony of the Sands.

US 50 through Nevada, dubbed the Loneliest Road in America, divides the state in half. On this asphalt counterpart to the old Pony Express Trail, manufactured attractions are the exception, but the occasional geologic curiosity does exist. One of these is Sand Mountain, 26 miles east of Fallon and reached by a short gravel road. A favorite site for off-road vehicle races, this 600-foot-high pile of sand is also famed for its ability to emit acoustical energy when disturbed, a phenomenon known as booming or singing sands.

Mysterious sounds have been the stuff of desert legend for centuries. In the year 1295, Marco Polo described them, during his journey through the Desert of Lop in China, as filling "the air with sounds of all kinds of musical instruments" and attributed them to "evil desert spirits." More recently, Charles Darwin, following a conversation with a resident of Copiapó, Chile, in 1835, related the noise phenomenon in his seminal text, *Voyages of the Beagle*, as "very surprising" and caused by "sand rolling down the declivity."

One of the more interesting aspects of this enigma is the wide range of reported sound. While "singing" is the generic descriptor, others abound: thunder, roaring, moaning, the buzzing of telegraph wires, bellowing, humming, drums, cannon fire, tambourines, and even the drone of low-flying aircraft. Visitors to Sand Mountain have described the noise as "the flight sound of a B-29 aircraft," and "a low, droning cadence—similar to a didgeridoo." Travelers as far away as 6 miles have reported hearing the sound, and others insist it has lasted for as long as fifteen minutes.

Sand Mountain is a seif dune complex that traces its history to Pleistocene-age mountain glaciers in the Sierra Nevada. When the ice melted, rivers carried sediment into an ancient lake to the east. The water evaporated slowly in a warming climate, exposing lake floor sand that was then blown to the northeast and deposited on the southwest flank of the Stillwater Range, where it resides today.

After more than a century of both field and laboratory studies, researchers are yet to close the book on the riddle of singing sands, but they have learned that certain physical characteristics are necessary before the advent of sand-song, whether the sands occur in North or South America, Asia, or Africa. Individual silica grains must measure between 0.011 and 0.015 inches in diameter, have a relatively high degree of roundness, be well to very well sorted by size, have moisture content of less than 1 percent, and have highly polished, seemingly frosted surfaces. These characteristics appear only when sand has been transported a certain critical distance and in the process has been reworked to the status of maturity, the state in which all the unstable minerals within sediment have been eliminated by weathering, leaving only quartz, which is the sole mineral of the Sand Mountain sands. Here is one case where old age has a competitive edge. Immature sands cannot hope for a singing career.

Finally, when the grains are loosely packed and lie high on the crest of a dune at a maximum angle of repose of around 34 degrees, a footstep, gravity, or wind can trigger an avalanche, producing a shearing effect within the moving mass. The downwind face of the dune then becomes a quivering musical instrument formed of millions of somersaulting granules, all vibrating in harmonic cooperation—euphonious sound with Mother Nature wielding the baton. Vive la unfinished symphony of the dunes.

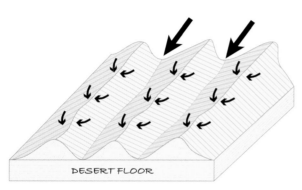

Seif dunes are long, narrow ridges oriented parallel to prevailing winds (large arrows). Secondary winds (small arrows) redistribute sand grains, producing the proper slope angle necessary for the orchestration of singing sands.

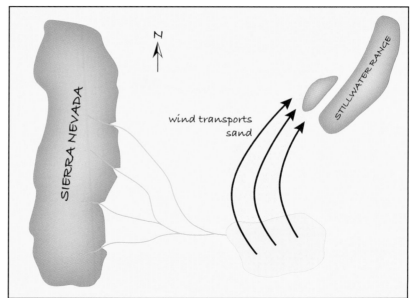

Born in the Sierra Nevada, sand was moved eastward via rivers to an ancient lake (blue). From there, wind transported the sand to its home beside the Stillwater Range. Along the way, reworking gave it the ability to eventually sing.

Sand Mountain, home to the rare gossamer-winged blue butterfly, is 2 to 4 miles long and 1 mile wide. —Courtesy of the Bureau of Land Management

51. Great Unconformity, Nevada

36° 11′ 55″ North, 115° 00′ 31″ West
Proterozoic Eon–Cambrian Period Unconformity

A worldwide gap in the rock record says Earth endured a maelstrom of change some 500 million years ago.

Every personal library should contain at least one good reference on history. If the criterion for choice is the biggest bang for the buck, then the selection should be a volume on geology. Generally $50 will buy a text covering 4,600 million years—or 92 million years' worth of narration per dollar spent. Quite a deal, but as always caveat emptor—let the buyer beware, because that history probably contains as many gaps as facts.

Recently, east of New Zealand, oceanographers discovered an area of the seafloor the size of the states of Alaska and Texas combined composed of basalt and almost barren of sediment. A sample of rock collected here would give no insight whatsoever as to what has taken place geologically in the time since the basalt was extruded approximately 60 million years ago. This finding is a paradigm of the pitfalls involved in writing Earth history: voids in time and history occur in every rock column.

Exactly 1 mile east of the intersection of Lake Mead Boulevard and Hollywood Boulevard in Las Vegas lies a classic terrestrial example. Here, at the base of Frenchman Mountain, rock inclined at 50 degrees abruptly changes from 525-million-year-old Tapeats sandstone to 1,700-million-year-old Vishnu schist. The age gap between the two rock types represents a full 25 percent of the history of the planet. Put another way, a one hundred–page history of Earth incorporating this Great Unconformity would contain twenty-five blank pages.

Unconformities exist worldwide and prompt a reaction among geologists akin to what librarians might feel about a book with missing or blank pages. They occur where erosion has stripped away one or more rock layers, or they represent a time period during which the production of new rock was interrupted. Defined simply as a substantial break, or gap, in the geologic record, unconformities are ubiquitous and occur in three basic formats. A disconformity is an erosion surface separating sequences of sedimentary rock with bedding planes that parallel each other. A nonconformity is an erosion surface separating either igneous or metamorphic rock from overlying sedimentary rock. And an angular unconformity is an erosion surface separating sequences of rocks with bedding planes that are not parallel.

Because Earth is subjected to cyclical periods of mountain building and sea level change, unconformities are continually being born: rising mountains erode while rising seas deposit sedimentary rock. As a result, it is estimated that many individual sequences of sedimentary rock represent at the very most 5 percent of Earth's history. Fortunately, however, the forces of erosion in one region are counterbalanced elsewhere by the products of deposition. By correlating these bits and pieces of the rock record in a comprehensive and chronological framework, the fundamental aspects of geologic history can be determined.

The Great Unconformity at Frenchman Mountain is a nonconformity composed of distinctive rock types. It can be traced from its exposure outside Las Vegas eastward into the inner depths of the Grand Canyon and beyond. It is labeled "great" because counterparts are common throughout the world wherever Precambrian rock is exposed, the Vishnu schist in this example. This universality supports the hypothesis that Earth underwent a prolonged period of change some 0.5 billion years ago. During the missing 1,175 million years represented by the Great Unconformity, bacteria and primitive forms of multicelled animals, whose bodies lacked hard parts, evolved to marine life and later terrestrial life supported by skeletons of chitin (a substance similar to the human fingernail), silica, and calcium carbonate. These Paleozoic life-forms were the ancestors of present-day life.

Because the missing pages attributed to the Great Unconformity can never be reclaimed, the exposure at Frenchman Mountain is a continuing reminder that much more of geologic history has been lost than is, or is likely ever to be, known.

Broken exposures of reddish Vishnu schist underlie light-colored beds of Tapeats sandstone that plunge to the left at a 50-degree angle to the horizontal. The trace of the Great Unconformity is shown as a dashed line.

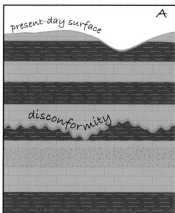
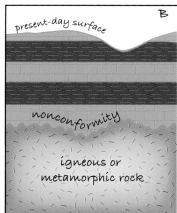
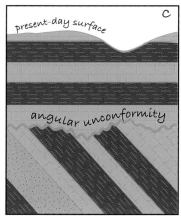

Disconformities (A) and angular unconformities (C) are both associated entirely with layered rock. In contrast, layered rock overlies either igneous or metamorphic material in a nonconformity (B).

The two rock types at Frenchman Mountain are easily recognized. Dark Vishnu schist (right) is composed of sediments that were metamorphosed. In contrast, the Tapeats sandstone is layered, crumbles easily, and contains marine fossils, evidence it was deposited along the shoreline of an ancient sea. Pen for scale.

52. Flume Gorge, New Hampshire

44° 05′ 44″ North, 71° 40′ 45″ West
Cenozoic Era Differential Weathering

A deep, dusky, and dank passageway separates igneous rocks with disparate weathering characteristics.

Every state has a nickname. Since colonial times, New Hampshire has been known as the Granite State, a less than unique moniker, since granite is not at all rare. Dig a hole deep enough on any plot of land in the world, and odds are pretty good granite will appear. Indeed, this durable, light-colored, coarse-grained rock—composed of two parts feldspar to one part quartz—is so common it is often called the "signature rock" of the continents. Why then does it represent the state of New Hampshire? The geologic map of New Hampshire's bedrock surface is liberally paved with exposures of granite. The youngest of these, the White Mountain Batholith, occupies a central geographic position.

This 150-million-year-old igneous mass has two claims to fame: the portion that has a distinctive red tone was quarried for a century for building and dimension stone, and it is the home to one of the region's most interesting geo-sites—Flume Gorge in Franconia Notch State Park. This oddity of topography, off exit 34A of I-93, is a revealing example of how the nuances of geologic process can reconfigure the normally subdued contours of a granite terrain. Granite is a plutonic rock, formed in deep-seated subterranean chambers by the slow cooling and crystallization of acidic magma. It appears on Earth's surface only after prolonged periods of uplift and exposure to the multiple forces of erosion have removed overlying layers of rock. Uplift encourages erosion, and rain, wind, running water, and glaciation are the forces of erosion.

By the end of the Paleozoic era, around 250 million years ago, the ancestral protocontinents of planet Earth were winding down their often-chaotic migration from one hemisphere to the other and coming together into a state of terrestrial adhesion. The supercontinent of Pangaea was born. However, just as a date of birth is step one along the path to death, this "one Earth" landmass partially girdled the globe for 50 million years before meeting its inevitable doom through fragmentation and dissection.

Beginning roughly 200 million years ago, deep-seated rotating heat cells, termed convection cells, began to swirl beneath Pangaea's underbelly, pulling the supercontinent apart by tug-of-war force. Continental-sized rifts developed, in the form of troughs bounded by faults that extended through the crust. This stretching and fracturing of the crust led to a state of reduced pressure in the mantle, creating conditions where solid rock turned to liquid and began to ascend, driven by buoyancy. In general, the less pressure on a mass of rock, the more likely it is to melt. As the molten slush approached the surface and began to condense to a solid state as granite, it cracked into segments separated by massive, vertical fractures.

Later, each fracture filled with a younger generation of magma that hardened as dark-colored tabular intrusions that were more susceptible to physical and chemical deterioration than the contiguous granite. At Flume Gorge, tenacious weathering and erosion have worn away one of these dikes to a depth of nearly 100 feet, creating a 12-to-20-foot-wide, 800-foot-long, precipitous, stream- and waterfall-filled passageway.

In 2003, a human profile sculpted by nature on nearby Cannon Mountain, known as the Old Man of the Mountain, collapsed from its perch. The very set of five fractures that had selectively maintained it for thousands of years finally yielded to the same destructive forces of weathering and erosion that created Flume Gorge. In their presence and in their comparison, these two locales exemplify an important lesson: nature giveth, and nature taketh away.

Flume Gorge (star) occurs in the western portion of the 40-mile-wide White Mountain Batholith, one of many granite intrusions (orange) that characterize half of the bedrock terrain of New Hampshire.

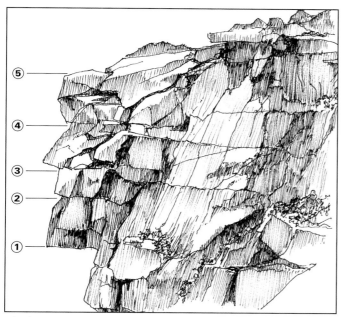

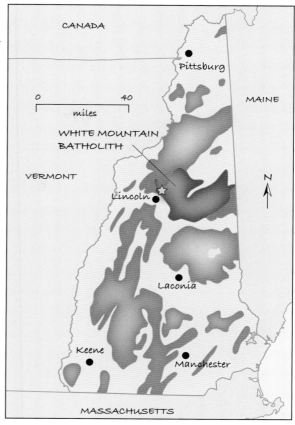

The fractures (1–5) that weathered in a peculiar way to form the profile of the Old Man of the Mountain also led to the destruction of this former rock star. —Courtesy of Patsy Faires, Kernersville, NC

Granite walls enclose 800-foot-long Flume Gorge, the void created by the weathering of a massive tabular igneous intrusion.
—Courtesy of Greg Keeler, Franconia Notch State Park

53. Palisades Sill, New Jersey

40° 51′ 14″ North, 73° 57′ 35″ West
Jurassic Period Magma Intrusion

An instance where a brutish, 1,000-foot-thick mass of molten mantle was injected into a subterranean venue.

Throughout geologic history certain events have significantly altered Earth, either physically or biologically. The tectonic reconfiguration that occurred at the end of the Paleozoic era, as studied along the eastern seaboard of the United States, is one example. During the late Carboniferous and early Permian periods, the southeastern border of proto–North America entered the final stages of colliding with proto–South America and proto-Africa. The tectonic event formed the Appalachian Mountains and squeezed the Iapetus Ocean that existed between them shut. The result was the birth of the supercontinent Pangaea. Once formed, this landmass—so intricately knit together—began to unravel. Europe, North and South America, and Africa drifted apart, creating a gulf that increasingly widened and filled with the waters of the developing Atlantic Ocean. This process continues today.

Evidence of this breakup appears on the New Jersey side of the Hudson River, along the massive Palisades cliff adjacent to Henry Hudson Drive in Palisades Interstate Park. This rock exposure is possibly, if not probably, the world's best example of a sill: a tabular igneous intrusion that parallels the bedding of the country rock it intruded. A sill is a geologic iteration of the slab of pastrami in a sandwich. Here is how it developed.

The breakup of Pangaea created a series of elongate grabens—troughs bordered on one or both sides by faults—from Nova Scotia to Florida. The graben that formed in the New York–New Jersey area is known as the Newark Basin. Over the course of 35 million years, the basin, 100 miles long by 50 miles wide, deepened and filled with more than 20,000 feet of debris eroded off the youthful Appalachian Mountains. Crafted from molten rock derived from the mantle of the Earth, the 1,000-foot-thick Palisades Sill intruded the lower portion of this rock column around 190 million years ago.

The massive sequence of sedimentary rock and interbedded igneous rock that filled the Newark Basin is known as the Newark Supergroup. The color of these rocks changes from bottom to top, reflecting the change in environmental conditions under which they were deposited. Alternating beds of siltstone and black shale beds of the Lockatong Formation, which were deposited within a large lake, comprise the base of the sequence. The alternating beds suggest the depth of this lake fluctuated on a cyclical basis. These dark-hued deposits contain fossilized bones and tracks of the earliest-known dinosaurs as well as evidence of the first mammals. The overlying beds are composed of reddish sandstone and shale of the Passaic Formation. The color results from the presence of hematite, an iron oxide that indicates the rocks formed in a semiarid environment. In short, the transition from the Lockatong to the Passaic formations suggests a change from a lake to a drier, warmer environment.

The best place to see the Palisades Sill is along Henry Hudson Drive, just north of the interstate approach to the George Washington Bridge. Within the geoscience community, this exposure is known as one of the most famous outcrops in America because of the clarity of its geologic features and its relationship to the breakup of Pangaea. Here, layered sedimentary rocks of the Lockatong Formation underlie the massive igneous structure of the base of the sill. A fine-grained, almost glassy, chilled contact was created when the magma came into direct contact with the cold Lockatong strata. The top of this sill, seen elsewhere, has the same chilled contact with overlying sedimentary strata, evidence the sill was intruded into preexisting rock. In short, here at the Palisades Sill, on full and dramatic display, is a geologic panorama born of the fragmentation of a once mighty supercontinent and the birthing of the now expansive Atlantic Ocean.

Close-up of the chilled contact between the Lockatong Formation (layered rock) and the intruded sill (top), located several hundred feet north of the George Washington Bridge. Rock hammer located on chilled contact.

The early stages in the breakup of Pangaea along an extensive rift zone (black line) created a chain of grabens along the east coast of the United States. The Newark Basin is one of these tectonic stretch marks.

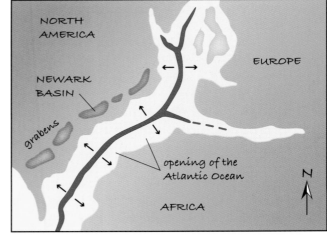

Cross section of the Newark Basin. Arrows denote relative movement along the fault.

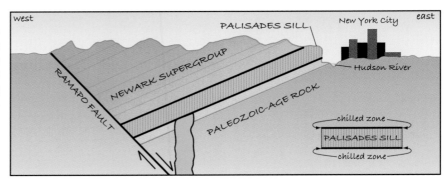

125

54. White Sands, New Mexico

32° 46′ 47″ North, 106° 10′ 20″ West
Quaternary Period Dune Development

A tableau of snowlike drifts of sand has an ancestry contrary to that of any favorite seaside beach.

The word *desert* is often misunderstood. It conjures up visions of heat and misery coupled with dryness, absence of vegetation, and acres of shifting sand. The Sahara of Africa and the ultradry Atacama of South America are good examples, but the snow-covered emptiness of Antarctica might be questioned. Since the term *desert* is defined simply as any region receiving less than 10 inches of precipitation per year, or an area where the evaporation rate at least doubles the precipitation rate, the Sahara, Atacama, and Antarctica all qualify.

The Chihuahuan Desert is another example. Straddling the Mexico–U.S. border, it is possibly the most biologically diverse of any desert in the world. Its northern extension, the Tularosa Valley, is home to White Sands National Monument, 15 miles southwest of Alamogordo. The titanium-white landscape of the monument is likened to a sea of sand and widely lauded as the largest gypsum dune field in the world.

The mountains bookending the Tularosa Valley—the San Andres to the west and the Sacramento to the east—came into being some 10 million years ago, when the axis of a gargantuan, north-south-trending anticline, one of the many topographic irregularities formed during the development of the Rocky Mountains, collapsed. Sedimentary rocks exposed in the fault-scarp borders of these mountains contain extensive volumes of gypsum, which had been deposited in a shallow sea that enveloped the region throughout the Permian period, tens of millions of years before the mountains collapsed. Rivers draining these mountain fronts dissolved this soluble mineral, carried it into the landlocked Tularosa Valley, and precipitated it there as selenite, a dark brown to yellowish, transparent, crystallized form of gypsum. Crystals up to 4 feet long ring the shoreline of Lake Lucero, a large playa that is the ghostly remnant of an even larger intermittent body of water—Lake Otero—that inundated the valley during the Pleistocene epoch.

Exposed to the vagaries of desert weathering, particularly those associated with extreme temperature fluctuations and precipitation cycles, the selenite crystals gradually disintegrate into sand-sized particles. When the prevailing southwest winds exceed 16 miles per hour, these sands begin to migrate and form dunes. At White Sands four dune shapes are constantly being constructed, depending on sand supply, amount of vegetation, and the strength and direction of the wind. The first to form are dome dunes, low, circular mounds that move downwind as much as 38 feet per year. Barchan dunes are crescent-shaped with tips that point downwind. They form under conditions with a limited supply of sand and strong, consistent winds. Parabolic dunes, with a form opposite that of barchan dunes, have upwind-directed arms partially anchored by plants. Long ridges of sand oriented at right angles to the prevailing wind are transverse dunes, which form when abundant supplies of sand are available.

Composed of 96 percent gypsum, the 115-square-mile sand field of White Sands National Monument (there are a total of 275 square miles of sand dunes in the region) is one of the world's most unusual natural wonders. When the wind is en force and maintenance vehicles are plowing a road through the dunes, the scene is reminiscent of a winter wonderland far afield from this high-desert showcase, where the abilities of wind to radically transfigure the landscape are on full display.

Curving and often coalescing ripple marks form when air currents swirl particles of gypsum into elongate ridges, wrinkling the surface of the dunes.

Simplified map of the White Sands region.

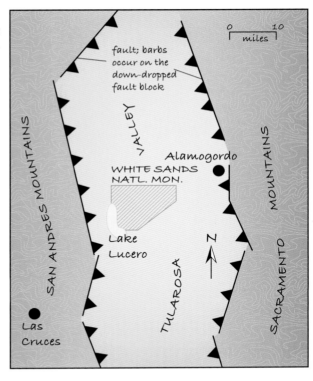

Barchan dunes are crescent shaped, with an upwind nose enveloped by downwind horns that encircle a central slip face. Wind-driven sand cascading down the slip face causes these dunes to move about 12 feet per year.

55. Carlsbad Caverns, New Mexico

32° 10′ 31″ North, 104° 26′ 41″ West
Cenozoic Era Cavern Development

Bizarre chambers superimposed on an oil field bear witness that not all caverns have a common heritage.

It is an almost unalterable law of nature that caverns form when sedimentary rock of the carbonate variety dissolves. There are nearly forty-five thousand caverns within the contiguous states alone, from the 390-mile-long, snakelike labyrinth of Mammoth Cave of Kentucky to Carlsbad Caverns in New Mexico, which could accommodate the U.S. Capitol building inside the 8.2 acres of its 370-foot-high Big Room, with 82 feet to spare up top.

The explanation of the process of dissolution, by which chemical weathering creates caverns, must of necessity make mention of the terms *carbonic acid*, *zone of aeration*, and *zone of saturation*. Carbonic acid is formed by the chemical union of the ordinary water of a summer shower and carbon dioxide, a natural constituent of the atmosphere. Worldwide, caverns originate within the zone of saturation, the subsurface region in which all available open space is filled with water. Here, carbonic acid–charged groundwater enhances fissures, joints, and bedding planes of soluble rock by slowly enlarging tiny voids to chambers of sometimes leviathan proportion. When the land is uplifted or the zone of saturation is lowered, these water-filled cavities empty. They then constitute the zone of aeration, rock composed of air-filled voids. If these voids are large enough, they can be explored by either spelunker or tourist—or even, as in New Mexico, house a cafeteria.

The three hundred or more caverns that underlie southeastern New Mexico and western Texas, including the 100 to 115 known caverns of Carlsbad Caverns National Park, however, originated via a different process. Here, the key component is sulfuric, rather than carbonic, acid. This story begins 260 million years ago, with the deposition of an immense thickness of limestone from the Delaware Sea, a shallow inland sea that covered present-day western Texas and southern New Mexico during the Permian period. Toward the end of the Cretaceous period, compressive forces associated with the Laramide orogeny—the Earth-altering tectonic episode responsible for the formation of the Rocky Mountains—uplifted the area. At that time, hydrogen sulfide–rich brines emanating from deep-seated deposits of petroleum and natural gas began to migrate vertically, following rock fractures. They then mixed with downward-migrating groundwater to form a highly toxic solution like the sulfuric acid in an automobile battery. This corrosive tour de force quickly and erratically consumed the limestone bedrock, leaving an interconnected array of fluid-filled caverns. Phase one of the Carlsbad story—cavern creation—was completed. Phase two—cavern ornamentation—followed.

As the land continued to rise, the level of the zone of saturation dropped and air replaced the fluid in the caverns. Droplets of downward-migrating carbonic acid deposited an infinitesimal amount of calcite on the ceilings of the voids, initiating the formation of icicle-like stalactites. When droplets fell to the cave floor, upward-growing stalagmites developed. Finally, as a fitting touch of decoration, flowstone, a sheetlike deposit precipitated from flowing, calcium-charged water, coated the cavern walls. Eventually, an almost incredible variety of speleothems—mineral deposits in caverns—created an enchanted grotto of indescribable beauty and splendor. The self-guided Big Room route winds 1 mile through this cavern unlike any other on the planet and offers a must-have opportunity to explore the ever-continuing cavern-making process.

Radiometric age dating of the stalactites and stalagmites in the Chinese Theatre shows they are 50 million years old and developing at a rate of 1 inch every 500 years. —Courtesy of the National Park Service

This triumvirate of stalagmites, the 62-foot-high Giant (left) and the Twin Domes, grows drip by drip as calcium-charged groundwater enters the Big Room and evaporates. —Courtesy of the National Park Service

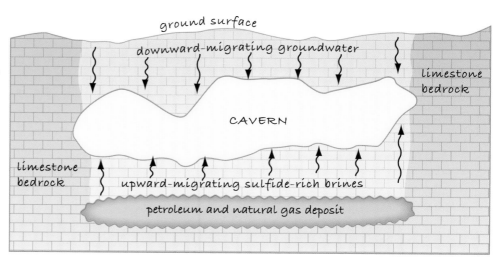

A toxic mixture of migrating groundwater and hydrogen sulfide–rich brines dissolved Carlsbad Caverns from limestone bedrock.

56. Ship Rock, New Mexico

36° 41' 24" North, 108° 50' 16" West
Paleogene Period Maar Eruption

The "winged rock" is the remnant of a feeder pipe that fueled a volcanic eruption of A-bomb magnitude.

A long-standing axiom of the real estate industry encourages the purchase of land because there's only so much of it, and it isn't being made anymore. Not true. The process of volcanism is continually creating new lands along the 40,000-mile-long oceanic ridge system at the rate of 1 square mile per year. Iceland is a good example. Totally volcanic in origin, this wannabe continent sits squarely astride the Mid-Atlantic Ridge that juxtaposes the North American and European tectonic plates. The periodic eruption of lava along this ridge increases the volume of Iceland by some 1,600 cubic feet each and every year—enough to fill a modern American guest room.

The most recognizable product of volcanism is, of course, the volcano. Today more than 1,500 volcanoes are categorized as "active," meaning they have recently erupted or may erupt again. Ten to twenty of them are churning away somewhere on Earth during the course of any year. Within the continental United States, evidence of recent volcanism is confined to areas west of the Mississippi River, principally along the Cascade Range of California, Oregon, and Washington and in the Four Corners region of the Southwest, where the Navajo Volcanic Field covers 7,700 square miles. This volcanic field is dotted with the remnants of dozens of once-active volcanoes, all sharing an age of 35 to 25 million years.

The most famous Four Corners example is Ship Rock, revered by the Navajo Nation as *Tsé Bit'a'í*, the "rock with wings," and acknowledged by geologists as an outstanding example of a diatreme—a breccia-filled volcanic vent born of a gaseous explosion. As impressive as it is today, accentuating the horizon of northwest New Mexico 10 miles southwest of the community of Shiprock, this 1,500-foot-wide, 1,700-foot-tall peak is a small remnant of a much larger Oligocene-age structure.

Around 30 million years ago, extension in the crust fractured the Four Corners region, a roiling disturbance much like what is happening today in Iceland, where uplift, rifting, and volcanic eruptions are creating new lands. As magma rose to the surface, decreased pressure released gases. The gas-charged slurry came into contact with near-surface groundwater, causing a blowout that formed a maar, a volcanic landform that looks much like a meteorite impact crater. Beneath the maar, fractured rock and magma solidified in the volcanic neck, the feeder system of the magma intrusion. Subsequent injections of less-gaseous magma parented three dominant, and several minor, basalt dikes that radiated outward from the maar like spokes from the hub of a wheel. Over tens of millions of years erosion reduced the landscape as much as 3,000 feet in elevation, destroyed the maar, and exhumed the diatreme—the breccia-filled volcanic throat that is now Ship Rock—as well as the radiating dikes.

The 5.5-mile-long South Dike, along with its shorter siblings, tracks straight as an arrow across the encircling desert floor from Ship Rock. The photographic attraction of these natural rock bulwarks would be even more impressive if the debris flanking their walls were removed. No matter, for even half buried in their own rocky garbage, they cast wide shadows across a landscape that is as much a scientific as a shamanic phenomenon.

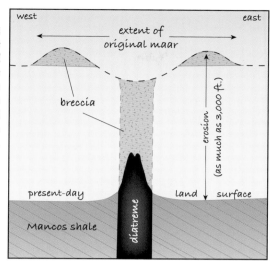

Cross section comparing the present-day land surface and the profile of the Ship Rock diatreme with what existed when volcanism was active.

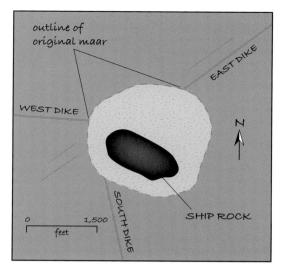

Three dikes radiate from Ship Rock. Erosion has greatly reduced the size and extent of the original volcanic structure.

Ship Rock is considered the most spectacular of the many exhumed volcanic features of the Four Corners terrain. —Courtesy of Farmington Convention and Visitors Bureau

South Dike is one of the largest and longest features of the Four Corners region. —Courtesy of Kirt Kempter, Santa Fe, NM

57. State Line Outcrop, New Mexico

32° 00′ 35″ North, 104° 29′ 54″ West
Permian Period Varve Development

This unprecedented 200,000-year catalog of climate change has inspired generations of geologists.

In communities across America, the inquiry "What's the weather like around here?" will elicit from the resident wag a variation on the same response: "Well, if you don't like it, in five minutes it will be different." Without doubt, atmospheric conditions have long been of foremost interest to humanity, probably even when humanoids first decided to venture beyond the protection of their caves. Today, to understand developing trends such as global warming, researchers collect massive volumes of historic data, much of it gleaned from ice cores extracted from the glacier fields of Greenland and Antarctica. At best these open a mere million-year window into the past. What of the other 4,599 million years of Earth's climate history?

A tantalizing answer can be found along US 62/180, 2 miles north of the Texas–New Mexico border, at the state line outcrop—a sedimentary rock exposure trumpeted as harboring one of the longest continuous climate records of the entire Phanerozoic eon. Even at first glance, its weathered surface appears unusual, and rightly so, because its pattern of alternating white and dark brown laminae—layers a fraction of an inch thick—constitutes a unique record. Each pair of light and dark laminae is a page from a thick book that documents 200,000 years of fluctuating climate conditions within an area formerly covered by an inland sea.

During the Permian period, an inland sea oscillated with regularity across a lagoon and open-shelf environment, depositing sediments that now give definition to the geography of present-day West Texas. The state line outcrop lies in the former corner of the Delaware Basin, one of several finger-shaped arms of this sea that extended into southeastern New Mexico. Its coastline to the west, north, and east was defined by reefs, and to the south by mountains.

As the curtain of time fell on the Paleozoic era, this basin began to fill with fine-grained layers of sediment, eventually becoming a thick sequence of sedimentary rock. Called the Castile Formation, it is composed of alternating laminae of dark brown calcite and lighter-colored units of anhydrite that extend for distances of more than 70 miles. Their repetitive characteristic suggests they are the product of a cyclical environment.

Geologists define each couplet of laminae as a varve, a unit of rock deposited within the span of one year. The calcite layers, stained by the organic remains of algae blooms, precipitated as the result of seasonal influxes of "freshening" volumes of water produced during extended periods of high relative humidity and cloudy skies. The thicker, lighter-colored anhydrite units supposedly formed during periods of lower humidity and sunny skies that encouraged the evaporation of the seawater, an increase in its salinity, and the precipitation of layers of anhydrite. The pattern of variable thickness is related to changes in these fluctuating weather conditions: the longer the periods of sunshine and lower humidity, the thicker the anhydrite layers.

Throughout the Castile Formation isolated bands of harmonic microfolds replace the otherwise horizontal banding. At one time, these contorted zones were believed to be either ripple marks or the result of the submarine slumping of sediments induced by earthquakes. A modern theory suggests they represent waves of tectonic compression that affected the region during a late phase of the development of the Rocky Mountains, some 50 to 40 million years ago.

The state line outcrop has enthralled legions of geologists. The twenty-millennia record of weather conditions contained within the horizontal and microfolded varves remains a depository of unparalleled significance in the annals of sedimentary rock history.

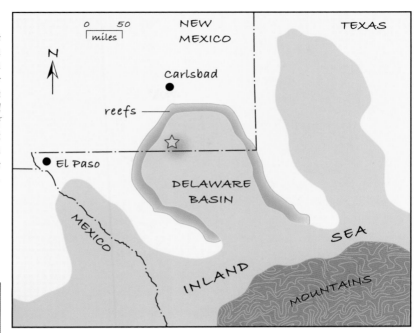

Map of the state line geo-site (star) region during the Permian period. Deepwater inlets of an inland sea (dark blue) separated mountains from a shelf of shallow water (light blue). Reefs (greenish blue) surrounded the Delaware Basin.

A count of the varves in this rock specimen indicates it records at least thirty years of paleoclimate history. Quarter for scale.

Examples of contorted bedding possibly created by compression forces rippling through the Delaware Basin long after the Castile Formation had been deposited. Quarter for scale.

58. American Falls, New York

43° 05′ 09″ North, 79° 04′ 07″ West
Quaternary Period Waterfall Retreat

This majestic derivative of the most recent ice age has a quest of self-destruction.

In the world of water in motion, waterfalls, also known as cataracts, enjoy exalted standing. Admirers argue little over definition: flowing water cascading across a perpendicular or sheer precipice. But what about bragging rights? Should "the greatest of all" claims be defined by distance from brink to foot or by flow volume? By either standard, Niagara Falls' two cataracts—Canada's Horseshoe Falls and American Falls—fail to attain significant worldwide status. Nonetheless, they are possibly the most-famous and most-visited waterfalls on Earth.

Approximately 12,000 years ago the 2.6-million-year reign of Pleistocene glaciation ended, a victim of global warming. As it retreated, the 2-mile-thick ice sheet left behind the Great Lakes—a system of meltwater surrounded by a blanket of drift. Freed from the pressure exerted by the ice—estimated at well in excess of 150 tons per square foot—the depressed bedrock rebounded, just as a ship rises when relieved of cargo. As the land rose, massive volumes of swirling meltwater separated into individual river systems. One of these, the ancestral Niagara River, connecting Glacial Lake Warren (today's Lake Erie) with Glacial Lake Iroquois (today's Lake Ontario), was no sooner born than it started to erode its valley downward to sea level. After cutting through the surface drift, the Niagara began to erode into a layer-cake, bedrock sequence of 450- to 425-million-year-old shale, sandstone, and limestone. Because of its greater resistance to erosion, the compact limestone soon stood out in relief in the form of a cuesta, an asymmetric ridge with a gentle slope on one side and a steep slope on the other. Water flowing over this asymmetrical ridge, called the Niagara Escarpment, was the birth of both of the cataracts forming Niagara Falls.

The shale beds underlying the surface rock at the falls are very susceptible to the erosive power of massive volumes of plunging water, a thunderous assault of splash and wave, continually undermining the limestone caprock and causing it to periodically collapse for lack of support. The effect is called cataract retreat, a process in which the falls migrate upstream as the bedrock beneath fails.

Since their birth along the Niagara Escarpment, at present-day Lewiston, New York, the falls have retreated 7 miles to the south at an average rate of 3 feet per year. If this recession rate were left unchecked, Niagara Falls would reach the mouth of Lake Erie in approximately 35,000 years. At that time, the catastrophic drainage of Lakes Erie, Huron, and Michigan, and soon thereafter Lake Superior—a volume equal to almost one-fifth of all the surface freshwater on Earth—would forever watermark the pages of geologic history. Fortunately, though, diverting water for hydroelectric generation has slowed the pace of retreat to a fraction of historic rates.

The accumulation of rock debris at the base of American Falls also slows down the rate of retreat, but its presence causes considerable damage to the acclaimed beauty of this honeymoon haven. In 1969, officials diverted the river for six months, reducing American Falls to a mere trickle. The intent of this operation was to remove most if not all of the rock debris, a task that would keep the falls from devolving to mere rapids. The estimated cost, coupled with public opinion, led to the decision to maintain the status quo and allow nature to take its course—one occasion when it was deemed wise to rewrite an old doggerel as "Even if it is broken, don't fix it."

Many rocks that lie at the base of American Falls are larger than the typical American family home. —Courtesy of Mike Smith, Niagara Tourism and Convention Corporation

Lateral retreat over the past six decades (hachures) is slower at American Falls than at its Canadian counterpart, the result of a massive buildup of eroded rock on the U.S. side that reduces the erosive power of the tumbling water. (Modified after Van Diver, 1985.)

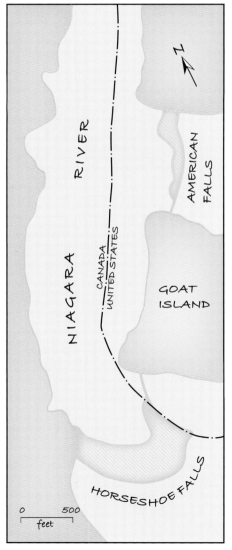

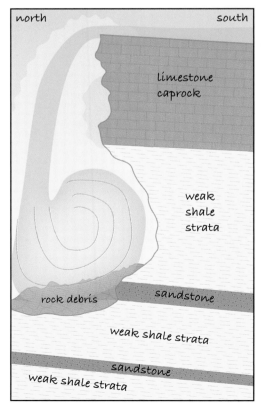

Cross section of American Falls. Waterfall turbulence continues to erode the weak shale, in effect undermining the limestone caprock.

59. Taconic Unconformity, New York

42° 14' 19" North, 73° 53' 11" West
Paleozoic Era Mountain Building

The construction plans of the Appalachian Mountains can be read at this acclaimed outcrop.

Mature in age and weather-beaten in appearance, the Appalachian Mountains have stood proud as a topographic definition of the eastern seaboard of North America for the better part of a quarter of a billion years. Devotees of these corrugated and faulted rocks are of two schools of scientific interest. The paleontology faction focuses on fossil content and the two seminal periods recognized as bookends to the Paleozoic era: the Cambrian explosion, when a wide variety of large, multicelled and shelled life-forms appeared abruptly on the geologic scene and the mass extinctions of the Permian period, that skull-and-crossbones moment when as much as 95 percent of oceanic species and around 70 percent of their terrestrial brethren disappeared. This paleontology contingent of investigators is laboratory and microscale oriented, using picks and awls to dissect a single slab of rock containing thousands of microfossils.

In contrast, the group primarily interested in rock composition, texture, and structure studies at the macroscale, using hammers and chisels to gather evidence from outcrops. When focused on those aspects of geology essential to understanding how the Appalachians developed, their analyses cross great distances—this venerable mountain chain extends 1,500 miles, from Newfoundland to Alabama—and involve hundreds of millions of years. Students of this orientation have long focused their attention on a singular, Rosetta stone outcrop in eastern New York State that encompasses evidence of all the key tectonic elements of the Appalachian birth. Called the Taconic unconformity, this outcrop is accessible outside the community of Catskill, 1 mile northwest of the intersection of US 9W and New York 23, along the northeast side of the off-ramp when traveling west along New York 23 to New York 23B.

This classic outcrop is constructed of two suites of strata. The older unit, the Ordovician-age Austin Glen Formation, composed of shale and siltstone, is intensely folded and stands almost on end. Overlying it are moderately disturbed Silurian- and Devonian-age carbonate beds of the Rondout and Manlius formations. This disparate juxtaposition of rock marks the character of this geo-site. Where they meet is the Taconic unconformity.

An unconformity is a generic type of rock relationship that represents a gap in the geologic record. The Taconic exposure is a special class, an angular unconformity, in which older, underlying rock is oriented at a steeper angle than younger, overlying strata. In this particular type of structure, sedimentary strata that were initially deposited horizontally are later affected by folding. Erosion nibbles away at the newly tilted strata before renewed deposition completes the picture.

Since both suites of sedimentary rocks at the Taconic site are tilted, more than one period of folding was obviously involved. In fact, these rocks were engaged in every one of the three orogenies—the Taconic, Acadian, and Alleghanian tectonic fender benders—that give definition to the Ordovician, Devonian, and Permian periods of geologic time (see the introduction for more information regarding these orogenies and time periods). While the evidence for the Taconic and Acadian orogenies can be observed in the rocks of this exposure, their adjustment during the Alleghanian orogeny is implied by regional studies conducted by "macroscale geologists" over a span of many decades.

By the end of the Permian period the eastern seaboard of the United States had been geographically configured by the Appalachian structure that prevails to the present. The Taconic unconformity shows how the evidence of the construction history of a mountain range, spanning hundreds of millions of years, can be compressed into a singular outcrop.

The sequence of events from the Ordovician to the Permian period that led to the creation of the Appalachian Mountains as interpreted at the Taconic unconformity outcrop. Mya = millions of years ago.

#		mya	
7		260	Alleghanian orogeny: Formations are further compressed and folded, and the Iapetus Ocean closes, during the final period of uplift of the Appalachian Mountains during Permian time.
6		400–350	Acadian orogeny: Formations are compressed, folded, and uplifted during Devonian and Mississippian time.
5		422–400	The Rondout and Manlius formations are deposited in the Kaskaskia Sea (another portion of the Iapetus Ocean), forming an angular unconformity during Silurian and Devonian time.
4		430	Erosion and planation of the Austin Glen shale during Silurian time.
3		470–440	Taconic orogeny: The Austin Glen shale is compressed and folded during Ordovician time.
2		465	The Austin Glen Formation (shale) is deposited in the Tippecanoe Sea, a shallow portion of the Iapetus Ocean, during Ordovician time.
1		480	Iapetus Ocean begins to close during Early Ordovician time.

West of Catskill nearly vertical shale and siltstone units (thin dashed lines show bedding) of Austin Glen Formation are distinctively separated from the steeply dipping beds of the Rondout and Manlius formations by an angular unconformity (dashed line).

60. Gilboa Forest, New York

42° 23′ 52″ North, 74° 26′ 50″ West
Devonian Period Fossilization

Only sandstone casts of stumps and roots remain of this first forest to shade the American landscape.

Eight thousand years ago forestland graced and shaded an estimated 45 percent of Earth's landmass. Today, due to the accumulative effects of human exploitation, a third of that percentage is gone. The result? A worldwide increase in land erosion, climate change, loss of animal habitat, and bothersome rates of extinction of tropical plants valued by the medical community.

It was a different story some 540 million years ago, as the Proterozoic eon of geologic time phased into the Paleozoic era. For more than 3.5 billion years after Earth had formed the land had lain denuded, lashed by wind and rain and sun, and unprotected by any form of canopy. Mother Nature had not yet added afforestation—the original greening of Earth—to her to-do list. However, over the next 160 million years she actively addressed the task, as evidenced by the discovery of fossilized plant remains, varying in size from inches to feet, in 380-million-year-old strata of Middle Devonian age.

The first documented evidence of a fossil tree stump in North America was made in 1850, when an amateur naturalist discovered a sandstone cast of a portion of a tree trunk in the valley of Schoharie Creek in southeastern New York. Eighteen years later, workmen blasting rock in a quarry near the town of Gilboa found several lumpy, vase-shaped, sandstone casts that measured 4 feet in height. A cast is a type of fossil that forms when sediment preserves the shape and some of the details of an organism. The sandstone casts formed as sand filled in cavities created by the decay of stumps and then later hardened to rock. The casts in the quarry rested in an in situ position and sported treelike characteristics. Expert analysis soon supported the suggestion they were indeed fossil tree trunks, documentation of the oldest-known forests to have existed on Earth. There was not enough evidence, however, to determine the overall form and shape of these woodland denizens.

It would be 136 years before the discovery of two different and spectacular fossil finds—one of a trunk and roots and the other a cluster of branches—would help researchers identify the earlier specimens as being a 380-million-year-old species, the "Adam" of all trees. Studies suggested that the fossil trees, named *Wattieza*, had been bottlebrush shaped, grew as tall as 40 feet to a crown of palmlike fronds, and reproduced by spore rather than seed. Vegetative organisms of this tree type greased the wheels of organic evolution, and a variety of trees—single-trunk, tall, woody plants—soon began to populate the landscape.

With this prototype firmly established, Devonian forest floors quickly became littered with branches, creating microenvironments favorable to the evolution of mites, centipedes, and spiders. Carbon-based soils formed, developing root systems held erosion at bay, and temperatures dropped as carbon dioxide was extracted from the atmosphere, opening the door for the evolution of broad-leaved forests.

These events transpired about the time proto–North America and proto-Europe were uniting and causing the Acadian orogeny, the second of three mountain building episodes that ultimately constructed the modern Appalachian Mountains. Both continents were then centered 15 degrees south of the equator, under a tropical climate. As the Acadian Mountains—which extended from Newfoundland to Georgia—grew in stature, a vast but shallow inland ocean, teeming with coral, trilobites, and jawed fish, extended to the west beyond present-day Indiana. Gradually, westerly flowing rivers eroded debris off the Acadians, turning the open waters into scattered regions of swamp and marsh. It was along one of these swampy zones that the Gilboa Forest came into being.

A display of the fossils can be found adjacent to the Gilboa Post Office and in the nearby Gilboa Museum, both located along New York 990V in Gilboa. The surrounding area is widely recognized as the most significant location in the world for the preservation of Devonian-age flora.

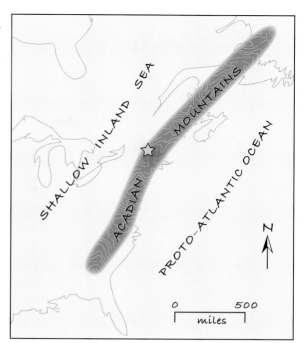

A tropical, shallow sea located west of the Acadian Mountains covered New York State during Devonian time. The Gilboa Forest (star) sheltered marshland along the shore of the sea.

This sandstone cast of a Gilboa Forest stump is believed to be the fossilized root remains of a primitive palmlike tree. The basal diameter is 3.5 feet.

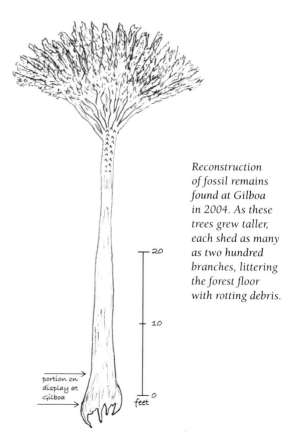

Reconstruction of fossil remains found at Gilboa in 2004. As these trees grew taller, each shed as many as two hundred branches, littering the forest floor with rotting debris.

61. Pilot Mountain, North Carolina

36° 20' 24" North, 80° 28' 28" West
Proterozoic Eon Mountain Building

A view onto the shores of the Iapetus Ocean is opened by the crown of a monadnock rising above the North Carolina Piedmont.

Halfway between Mount Airy and Winston-Salem, just around one of the gradual bends of US 52, the dominating profile of Pilot Mountain suddenly looms at 2,421 feet above sea level, a rock sentinel standing guard over the nearly flat topography of the North Carolina Piedmont. Some geologists call the feature an inselberg, a residual mountain rising abruptly from an extensive lowland surface in a hot, dry region. Given the prevailing weather conditions of the region, however, a more appropriate label is monadnock, an isolated mountain rising conspicuously above a peneplain in a temperate climate.

The Piedmont, French for "foot of the mountains," is the largest physiographic province in the Tar Heel State. It exhibits a rich geologic history that frequently presents the field geologist with more questions than answers. Throughout Paleozoic time, as part of the long-term geologic construction of the Appalachian Mountains, fragments of ancient landmasses crunched into proto–North America and were welded to it, creating a crazy quilt of exotic terranes—masses of metamorphic and igneous rock bounded by major fault systems. One such assemblage is the Sauratown Mountains anticlinorium, a structure composed of a series of folded fault blocks composed of tabular masses of rock stacked one upon the other and oriented convex upward. The anticlinorium is bounded to the south by the Carolina terrane and to the north by the Blue Ridge terrane and the Blue Ridge Mountains.

Atop Pilot Mountain is the much photographed rock star of this anticlinorium: Big Pinnacle, a massive, 200-foot-high, circular ledge. Here, a silica-rich metamorphosed sandstone (quartzite) unit of the fossil-barren Sauratown Formation grades downward into interbedded units of metamorphosed shale (slate) and quartzite. These 550-million-year-old Sauratown rocks overlie a thick sequence of gneiss dated to 1,200 million years of age, a radiometric age that probably represents the timing of metamorphism rather than deposition of the original sedimentary material.

The Sauratown Formation was deposited in the shallow waters of the developing Iapetus Ocean following the breakup of the supercontinent Rodinia, some 750 million years ago. Those sediments farther offshore were fine grained and variable in mineralogy, eventually becoming shale. Shoreward, into the regime of decreasing water depth and higher energy, sand-sized sediment bordered a surf-intensive beach not unlike that of the present-day east coast of Florida. Crossbedding and ripple marks found within the quartzite rock wall along the 0.8-mile-long Jomeokee Trail, which encircles Big Pinnacle, provide evidence for this nearshore environment, where water currents disturbed the sand.

As the continental plates that today constitute Africa, Europe, and North America merged to form the supercontinent Pangaea, the Iapetus closed. The tectonic pressure folded, faulted, and compressed the Sauratown rocks into metamorphic equivalents—shale to slate and sandstone to quartzite—and formed the Sauratown Mountains anticlinorium, rearranging the once simple, layer-cake rock sequence into a rolling complex of elevated and stacked anticlines. Millions of years of erosion have altered this contorted scene to today's monadnock.

In recent history, Pilot Mountain was a principle landmark for early settlers moving through the area. The native Saura Indians knew it as Jomeokee, the "Great Guide" or "Pilot." It is but one of several prominent monadnocks that interrupt the low relief of the Piedmont Province in North Carolina, among them Stone Mountain (Wilkes County), Crowders Mountain (Gaston County), and Hanging Rock (Stokes County).

Big Pinnacle, the dominant topographic feature at Pilot Mountain, juts a full 1,400 feet above the flat terrain of the Piedmont.

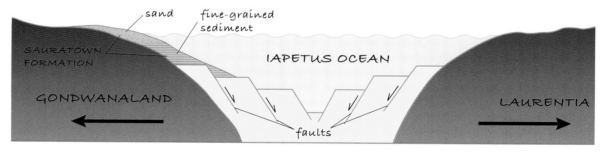

Interpretation of the shoreline environment in which the Sauratown Formation was deposited following the breakup of Rodinia into the landmasses of Laurentia and Gondwanaland. Arrows denote relative movement of landmasses and movement along faults.

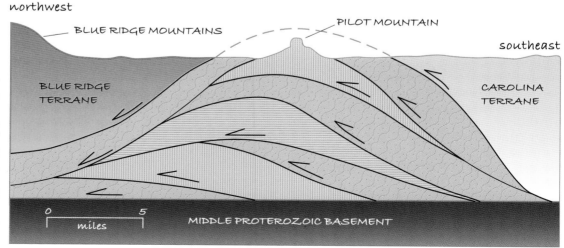

Pilot Mountain lies near the crest of the Sauratown Mountains anticlinorium, a convex-upward structure composed of a series of Proterozoic-age folded masses of rock (green and orange) separated by faults (black lines) and stacked one upon the other. Arrows denote relative movement along faults.

62. South Killdeer Mountain, North Dakota

47° 26′ 46″ North, 102° 52′ 57″ West
Cenozoic Era Topographic Inversion

An unusual interplay of volcanic activity, uplift, and erosion has in effect turned a lake inside out.

Anyone who studies the lay of the land of the contiguous United States might unconsciously imagine the surface of a pool table when North Dakota is mentioned. This Great Plains sector of space and open sky is the very epitome of a prairie environment. Travel brochures properly describe distant horizons and boundless vistas, rather than looming mountains and alpine scenery. Nevertheless, the landform lexicon of North Dakota recognizes at least eight "mountains," most no more than accumulations of glacial drift that seldom exceed 150 feet in height. South Killdeer Mountain, northwest of Killdeer and adjacent to the badland topography of west-central Dunn County, is an exception. This sentinel of flat-lying rock never felt the scour of a continental glacier or the impact of tectonic turmoil. It's a mountain of a different geologic stripe—one that essentially has been turned inside out.

Western North Dakota is a semiarid land of cold winters and unpredictable summers, quite a change from the tropical environment of 60 million years ago, during the Paleogene period, when swamps dominated the landscape. Shaded by forests of lofty cypresses and low-lying ferns, the aquatic lowlands were home to life-forms ranging from palm-sized snails and clams to a robust crocodile-like reptile called *Champsosaurus*.

Because global cooling was under way, glaciers already had begun to take root in distant Antarctica, a harbinger of the ice age that would reach maturity a few million years later. As Paleogene time aged to Neogene, shale, siltstone, and low-grade lignite coal deposits gave way to layers of sand and gravel in Dunn County. Fetid swampland evolved to oxygenated lakes engulfed by savannah plains and flushed by numerous rivers. Periodically, clouds of ash that blew in from erupting volcanoes in Montana and Wyoming filtered the sunshine.

Within this changing environment, four formations were deposited in a layer-cake arrangement. Three of these—the Sentinel Butte, Golden Valley, and Chadron—merely hold up South Killdeer Mountain. Formed of volcanic ash, the fourth formation, the Arikaree, is the very reason the mountain exists. The volume of ash that fell on dry land compacted over time into ordinary sandstone and siltstone, while the ash that settled into the many lakes that dotted the landscape interacted with freshwater, which as a result precipitated a compact caprock. This chemical precipitation occurred in much the same way that gravel interacts with cement to form concrete. The distribution of the caprock exactly mirrored the surface area of the lakes in which it was chemically created.

About 5 million years ago, a period of uplift invigorated the region, exposing it to renewed forces of erosion. Since the ash-bearing rock that formed ashore eroded at a faster rate than the dense caprock, lowland lake became highland butte, and South Killdeer Mountain was born—a classic example of inverted topography.

Southwestern North Dakota is peppered with buttes that formed the same way. South Killdeer Mountain, as well as an outlier located to the south, is the most recognizable among these because of the presence of a well-exposed crown of erosion-resistant Arikaree caprock. The perspective from the top is geologically akin to the view from the bottom of a 25-million-year-old fossil lake—a mountain turned inside out. Or, is it a lake turned outside in?

The outlier immediately south of South Killdeer Mountain is prominently crowned by a caprock of Arikaree Formation. —Courtesy of Edward C. Murphy, North Dakota Geological Survey

Different rates of erosion have transformed the former low-lying lake terrain of the South Killdeer Mountain area (left) into a prominent ridge and outlier crowned by a dense caprock.

Close-up of the erosion-resistant Arikaree caprock. —Courtesy of Edward C. Murphy, North Dakota Geological Survey

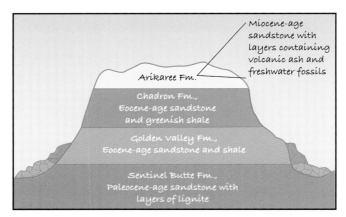

Four layer-cake formations make up the geology of this geo-site.

63. Hueston Woods, Ohio

39° 35′ 34″ North, 84° 46′ 12″ West
Ordovician Period Fossilization

Hundreds of definitive fossil species give evidence of a tropical sea that once covered North America.

Geology is a field of study of unusual diversity—an amalgamation of the core sciences of biology, chemistry, physics, and mathematics with many subbranches. One of the most popular of these is paleontology, the study of life in past geologic time as known through the fossil record. The history of paleontology is replete with the names of pioneers who elevated the scientific worth of fossils from the commonplace to the enlightening. Thomas Jefferson collected mammoth bones, Mary and Louis Leakey crafted new versions of human genealogies, and Donald Johanson discovered the bipedal hominid Lucy.

Many chapters in the book of life remain to be written by the amateur of today—the professional of tomorrow—who will by convention become initially acquainted with generic fossils in the wild and then move on to a career investigating specialized types. In that pursuit, Hueston Woods State Park, 7 miles north of Oxford, is an optimal place to start. The limestone and shale strata accessible here in both streambed and roadcut are world famous for the abundance, diversity, and museum-quality appearance of their invertebrate fossils.

During the waning centuries of the Ordovician period, the Buckeye State was positioned 20 degrees south of the equator and drenched by the tropical waters of the Tippecanoe Sea, classed as the greatest-ever submergence of any continental landmass. Mud was the deposit of choice offshore, while soupy and slimy ooze, sediment composed of at least 30 percent calcareous skeletal remains, accumulated in clearer inshore waters. This warm, oxygenated sea created a utopian environment for marine organisms. In life they flourished—in death they were preserved en masse. The Ohio Division of Geological Survey has suggested that if all the fossils were removed from the Ordovician strata beneath Cincinnati, the city would slump to below sea level.

Over seven hundred species of Ordovician life have been recorded in the strata of southwestern Ohio. While many are rare, the casual weekend explorer can easily identify in the time span of an hour a half dozen common species within the confines of Hueston Woods State Park. (Because Ordovician fossils are so abundant in the park, individuals are permitted to collect specimens for personal collections. Park personnel will happily recommend collecting areas.) The most abundant are bryozoans (moss animals), found as encrusting, twig- and fan-shaped colonies of microscopic individuals. The equally common brachiopod lived alone, attached to the seafloor within a shell of bilateral symmetry. Next in abundance are fossils of the mollusk family, composed of pelecypods (clams), cephalopods (squid), and gastropods (snails), all 1 inch or so in size. The solitary, extinct, several-inch-long horn coral is a favorite of amateur paleontologists because of its size, wrinkled surface, and resemblance to a bull's horn. Crinoids, also known as sea lilies, are related to the starfish, sand dollar, and sea cucumber and have beautiful, perfectly developed fivefold symmetry, a characteristic of this whole group of marine life.

Beyond any doubt, the most highly prized Hueston Woods fossils are trilobites, extinct relatives of today's lobsters and crabs, but also like the earthworm because they fed by ingesting seafloor mud. The animal takes its name from its unique shape, formed by two grooves that divided the body into three separate areas—thus "tri-lobe." In 1985 a 14-by-10-inch specimen of *Isotelus* became the official state invertebrate fossil of Ohio.

A visit to Hueston Woods State Park, even a short one, is proof positive of why specimens of Ohio Ordovician fossils are proudly and prominently displayed in museum collections throughout the world.

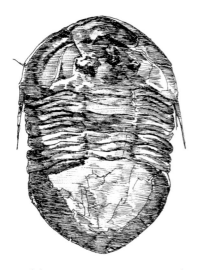

Trilobites were marine organisms that matured by periodically molting their external skeleton, a hard, flexible shell made of material similar to that of the human fingernail —Drawing by Patsy Faires, Kernersville, NC

Horn corals often grew to lengths of 6 inches, double the size of this specimen, which is stained red for clarification. Six-inch ruler for scale. —Courtesy of Charles Herron, Wilmington, NC

At least four species of brachiopod encrust this slab of limestone, which appears to be a rock composed completely of fossils. Six-inch ruler for scale.
—Courtesy of Charles Herron, Wilmington, NC

64. Big Rock, Ohio

40° 17' 14" North, 84° 08' 05" West
Quaternary Period Continental Glaciation

Analysis of this huge erratic verifies its place of birth and mode of transportation to its lasting location.

They exist in haphazard distribution on every continent and come in all sizes, but those subject to "rock obesity" receive most of the publicity. They have been labeled wanderers, orphans, aliens, and even immigrants. Their presence has been attributed to floods, icebergs, extraterrestrial transit, and subterranean emergence. A few are sedimentary in composition, but metamorphic and igneous types are most common. West of the Mississippi River they are confined to higher elevations, while none have been found throughout that portion of the United States south and east of a line connecting the courses of the Ohio and Missouri rivers with the Mason-Dixon Line. In short these rocks are, in a geologic sense, erratic—but not without explanation.

Consider the rounded, 1,250-cubic-foot, 103-ton specimen lying sheltered in Tawawa Park on the east side of the community of Sidney, in Shelby County. Laborers laying railroad track originally discovered Big Rock in a dense forest, but it was 1878 before a geologist reported it to be the "largest bowlder" yet observed in the state. A garden variety of granite, Big Rock lies enmeshed in 100 feet of 23,000-year-old ice age soil that overlies limestone and dolomite bedrock deposited during the Silurian period. Its size, shape, isolation, and starkness beg inquiry as to what it is, how it got to where it is, and its origin. Finding the answers is an interesting exercise in geologic analysis.

Ohio is a land of Paleozoic-age sandstone, shale, and limestone overlain by layers of rock debris deposited by glaciers. When large boulders are discovered within this field of debris, geologists immediately begin to posit that they are glacial erratics. Of varying size, beach ball to boxcar, these imposing rocks are born when glacial meltwater penetrates bedrock fractures, freezes, and then expands, breaking rock fragments from the parent formation that the glacial ice then plucks from the landscape and incorporates and transports in its frozen mass. When the ice melts, having reached its farthest point of advancement, and the glacial front retreats, the quarried remnants are left behind, irrefutable testimony to the awesome power of glaciation.

The route taken by advancing ice is a clue to the origin of any glacial erratic. Just as a rapidly braking automobile produces skid marks and a snail leaves a glistening trail of slime, glaciers leave evidence of their travels. Rock debris incorporated in the base of a flowing mass of ice sandpapers the bedrock over which it moves, forming striations, thin grooves incised parallel to the direction ice flowed. In addition, ice picks up the mineral character of the bedrock over which it travels and carries this distinctive chemical fingerprint downstream. An upstream plot of both striations and mineral trace will, in theory, lead directly back to the erratic's point of origin. Such a plot indicates Big Rock originated within a portion of the 985-million-year-old (Proterozoic age) Grenville Province of Ontario, some 150 miles north of Toronto.

The analysis is now complete. Quarried from the ancient basement rock of southeastern Canada by glacial ice, transported southwest and south some 700 miles and unceremoniously deposited in Shelby County toward the end of the most recent ice age, Big Rock remains today intriguing and demonstrative evidence of the mile-thick ice sheet that once enveloped the state in its chilling grip. A thorough postmortem review of the facts has demystified another enigma of nature.

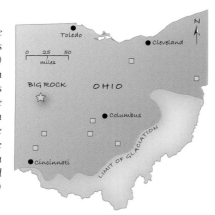

All eight of the well-known erratics (squares and star) in Ohio lie within the area that was glaciated during the Pleistocene epoch (green). Erratics are absent within the nonglaciated area (blue). (Modified after Hansen, 1984.)

Glacial striations (dashed lines) and mineral-trace analyses suggest the Big Rock (star) was plucked from exposures of Proterozoic-age granite in Canada and finally laid to rest in Shelby County.

The mineral make-up of Big Rock has no association at all with the bedrock composition of western Ohio. This is one of the key characteristics that delineates an erratic from any other large rock.

65. Kelleys Island, Ohio

41° 36′ 58″ North, 82° 42′ 26″ West
Pleistocene Epoch Glaciation

Sculpted by ice, the specifics as to how these grooves were made is still rife with controversy.

The science of geology was birthed on the backs and in the minds of many academics, who quickly discovered their logic ran counter to established wisdom. Hidebound resistance was especially prevalent early on, when the tenets of theology largely governed the interpretation of natural law. Glaciation is a case in point. In 1760 the Swiss geologist Horace-Bénédict de Saussure suggested the lineations chiseled into bedrock surfaces miles downslope from the snouts of alpine ice fields proved glaciers had once extended beyond their extant position. He was roundly ignored.

Resistance to the idea of ice in motion continued for another eighty years until de Saussure's countryman Louis Agassiz published his classic treatise *Studies on Glaciers*, in which he listed evidence for a great ice age that had once draped Switzerland with a frozen landscape reminiscent of that enveloping Greenland. At about the same time, scientists scrutinizing the gouges and striations that ubiquitously cover the Paleozoic rock surfaces in Ohio focused on a series of gargantuan grooves that are etched like undulating waves into the limestone face of Kelleys Island, in Lake Erie.

These megafurrows have garnered widespread attention since their discovery in the 1830s. Quarrying activities destroyed the largest of them during the early 1900s. After this industry temporarily ceased operations in 1932, the state acted to protect a small section of the grooves. Several decades later, controlled excavation of the overlying glacial soil revealed a beautifully preserved, several-hundred-foot-long extension of the already exhumed grooves. Billed as the most spectacular of their genre in the world, these exposures are on view today in the Glacial Grooves State Memorial, at the terminus of Division Street on Kelleys Island.

The surface of these deep first-order grooves, tattooed by shallow, second-order, linear scratches, provides insight into their origin. There is little doubt the motion of a great mass of ice that was contaminated with multitudes of gritty rock formed the minutely inscribed, generally parallel, needle- to inch-width lines. Just as sandpaper scratches the surface of a plank of wood, rock-laden glaciers scratch the surface over which they move, leaving striations that are a long-lasting record of the direction of ice movement. Aside from these surficial cosmetics, however, the deep first-order grooves themselves remain the subject of continuing controversy. Glacially formed, yes, but what are the specifics?

Measuring on the order of 35 feet wide and 15 feet deep, and extending for nearly 430 feet, these monster channels are sometimes linear, occasionally sinuous, commonly undercut, and always gargantuan in proportion to kindred 18,000- to 12,000-year-old features located elsewhere throughout the formerly glaciated terrain of the United States. Several different theories have been presented regarding their origin:

(1) Boulder-riddled ice further scoured the base of preexisting exposures of the east-dipping bedrock.

(2) Sediment-laden meltwater flowing beneath the ice sheet eroded them.

(3) They are meltwater channels that were eroded out of bedrock beyond the margin of the ice sheet.

(4) Rock-charged ice deepened and widened small, preexisting river channels.

While their precise genesis remains unresolved, the presence of these not-found-elsewhere ice age furrows is a lasting and visually significant monument to the power of moving ice when it makes contact with Earth's hardcore bedrock.

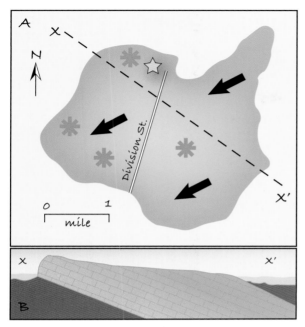

Map (A) and cross section (B) of Kelleys Island showing the direction of ice (arrows), location of the quarries (asterisks), and Glacial Grooves State Memorial (star).

Second-order striations vary from those barely visible to some 1 inch or more in width. White pencil for scale. —Courtesy of Mike Monnett, Kelleys Island State Park

The main attractions at Kelleys Island are the first-order, big-boy grooves deeply scoured into the limestone bedrock at Glacial Grooves State Memorial. —Courtesy of Mike Monnett, Kelleys Island State Park

66. Interstate 35 Roadcut, Oklahoma

34° 21' 04" North, 97° 08' 56" West
Paleozoic Era Mountain Building

A roadcut reflects varied geologic, industrial, and scientific events of the Arbuckle Mountains.

At the onset of the Paleozoic era the world was radically different. The continents were smaller and barren of plant and animal life. Because Earth twirled faster, the days were shorter and there were more than 365 per year. Proto–North America was little more than a fair-sized landmass straddling the equator, bathed by a warm, shallow sea that teemed with evolving species. A handful of bowl-shaped depressions, similar to the modern-day Hudson Bay, were beginning to alter the low-relief topography. One of these, the Arbuckle Basin, has played a major role in the economic and industrial development of the Sooner State. This story of rock and resource appears along a 7-mile stretch of I-35 that slices through the Arbuckle Mountains between mile markers 44 and 51, north of Ardmore.

As the Cambrian period dawned in Oklahoma, advancing tropical waters sorted the existing blanket of sediment into like-sized suites that would later become sandstone and shale. Without nearby highlands to act as a source of additional sediment, the only new deposition in the sea was the precipitation of calcium carbonate. Throughout the following 200 million years, the slowly sinking Arbuckle Basin filled with limestone and dolomite, which eventually exceeded 15,000 feet in thickness. This period of tectonic quiescence ended during Pennsylvanian time, when limestone deposition gave way to that of shale and sandstone, presaging the birth of the Rocky Mountains. The entire flat-lying sedimentary column was then folded into a gigantic anticline and faulted, thereby thrusting the igneous basement rock upward. The Arbuckle Mountains were born.

Between mile markers 44 and 47, rocks tilted at 40 degrees form the south limb of the Arbuckle anticline, the geologic core of the mountains. After a heavy rainstorm washes away weathered sediment, the extensive limestone outcrops along the first mile (mile markers 44 to 45) yield outstanding specimens of brachiopods (lampshells), gastropods (snails), bryozoans (moss animals), crinoids (sea lilies),

and the occasional museum-quality trilobite. Golf ball–sized carbonate concretions, hard and compact masses of mineral matter, weather out of the black shale immediately north of mile marker 44. The nuclei of these concretions, around which the mineral matter precipitated from groundwater, are believed to be minute fossil particles of feces excreted on the ocean floor by Devonian-age sharks.

The limestone strata exposed along the 0.5 mile leading to mile marker 45 contain layers of sandstone that represent important hydrocarbon reservoirs throughout the state. The 1928 discovery of oil within these Ordovician rocks quickly became a two-edged sword: Oklahoma City grew into a boomtown, but prolific production caused the market value of petroleum to plummet from $1.56 to $0.15 per barrel. Geologists estimate the producing zones composing this 1,658-acre bonanza, which is called the Oklahoma City Oil Field and touted as one of the greatest oil discoveries in history, will ultimately yield 1.2 billion barrels of oil.

Seven years before the discovery of the Oklahoma City Oil Field, this same sequence of rock had become the site of the first successful testing of seismic reflection, a process in which geologists create and evaluate energy waves in order to define subsurface rock structures. For nine decades this process has been the most successful means employed, onshore and offshore, to find and develop new reserves of oil and natural gas throughout the world.

The crest of the Arbuckle anticline is marked by the exposure of near-horizontal limestone beds in the vicinity of mile marker 48. One mile farther north, limestone strata plunging at a 40-to-50-degree angle to the north form the north limb of the anticline.

Hundreds of professional and student geologists visit this roadcut annually, and for good reason. These strata are not only superb exposures of folded and faulted Paleozoic rocks but are also the oldest rocks found at the surface anywhere between the southern Appalachians and Rocky Mountains.

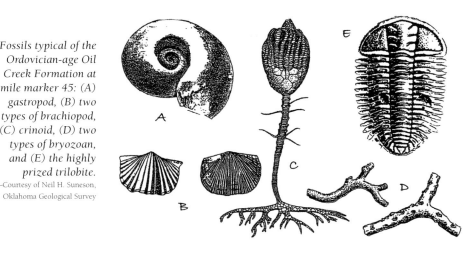

Fossils typical of the Ordovician-age Oil Creek Formation at mile marker 45: (A) gastropod, (B) two types of brachiopod, (C) crinoid, (D) two types of bryozoan, and (E) the highly prized trilobite.
—Courtesy of Neil H. Suneson, Oklahoma Geological Survey

South-plunging (to the left) limestone and cherty shale units of the Mississippian-age Sycamore Formation at mile marker 44.
—Courtesy of Richard D. Andrews, Oklahoma Geological Survey

67. Mount Mazama, Oregon

42° 56′ 02″ North, 122° 06′ 09″ West
Quaternary Period Volcanic Activity

This Ring of Fire site is home to an active volcano and the deepest lake in the United States.

Each of the 1,500 active volcanoes that pockmark the land surface of the world has experienced some type of eruption within the past 10,000 years. Approximately half identify the Ring of Fire, the zone of volcanic and seismic activity that encircles the basin of the Pacific Ocean. The western and northern segments of the ring are composed of more than forty volcanoes that make up the nation of Japan and the more or less fifty-seven that partially identify the Aleutian Islands. Closer to home, the thirteen towering cones associated with the High Cascades of the Pacific Northwest form a major portion of the eastern segment.

One High Cascades volcano stands apart, isolated by topographic profile and history from its geologic siblings. Constructed of alternating layers of lava and pyroclastic rock, this 8,151-foot-high caldera—a collapsed volcanic cone—is evidence of a modest igneous birth, followed by pulses of growth that progressed erratically until interrupted by catastrophic death, only to be reincarnated to second-generation status. Catalogued as Mount Mazama, it is familiarly known as Crater Lake, a name derived from the association of the vent with the steep-walled shoreline of the deepest body of inland water—1,958 feet at its deepest point—in the United States. Today, this Oregon landmark is the central attraction in Crater Lake National Park, accessed from both the north and south by Oregon 138 and Oregon 62. The view along the encircling, 33-mile, paved Rim Drive presents an informative perspective on the geologic complexion of both the volcano and the lake.

The embryonic form of Mount Mazama arose 420,000 years ago with roots firmly anchored in the north-south-oriented belt of late Neogene–age lava flows that comprises the basement rock of the High Cascades. Throughout an extended period, a series of eruptions alternating between quiet emission and explosive discharge of molten material built a volcanic cone that grew in physical complexity and size, reaching an estimated height of 12,000 feet. Its flanks soon became etched and scoured by glaciers constructed from moisture-bearing winds blowing off the Pacific Ocean.

Once Mazama had reached maturity, volcanic activity ceased and a verdant forest replaced its glaze of retreating ice. It seemed destined for a quiet old age, but then a shift in chemistry within the magma-filled furnace feeding Mazama produced a viscous, gas-charged mixture of slush that slowly ascended into the volcanic plumbing system. A one-two punch was about to be delivered.

The first blow occurred roughly 7,700 years ago, when thunderous explosions slashed through the flanks, ejecting an incandescent, 12-cubic-mile volume of ash skyward at twice the velocity of sound. Within days, winds distributed this dust plume over a multistate area, and the stage was set for the knockout punch. A geologic eyeblink later, 5,000 feet of the volcano's crown—15 cubic miles of rock—sank into the now partially voided subterranean magma chamber. Decapitation by collapse instantaneously transformed Mazama into a 6-mile-wide, 4,000-foot-deep caldera.

Although centuries of precipitation have since filled the colossal caldera, this story continues. About 1,000 years ago resurgent eruptions built Wizard Island, the son of Mazama that today reaches 760 feet above the spectacularly clear and reflective blue waters. Will this wounded and routed giant erupt again? Don't bet against it. Remember, nearby Mount St. Helens was once considered extinct.

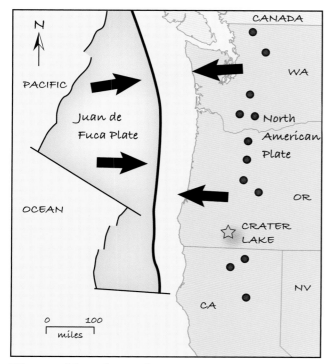

Volcanoes of the High Cascades (dots) trace their origin to the ongoing collision of two major tectonic plates, which forms deep-seated magma chambers that sporadically erupt volumes of lava and ash. Arrows denote direction of plate movement.

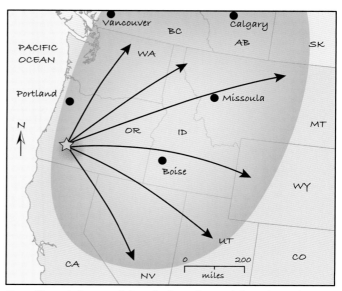

Prevailing winds carried the ash plume (orange) from the explosion of Mount Mazama (star) across eight states and three provinces.

Following the spectacular collapse of Mount Mazama, Wizard Island initiated a chapter of reincarnation.
—Courtesy of National Park Service, Crater Lake National Museum and Archive Collections

68. Lava River Cave, Oregon

43° 53' 42" North, 121° 22' 11" West
Pleistocene Epoch Volcanic Activity

A mile-long, glazed subterranean passageway bored by swirling currents of 2,000-degree molten rock.

Other than as the Beaver State, how might Oregon be characterized? The Exploration State, since it was the wintering home for the groundbreaking Lewis and Clark Expedition? The Pioneer State, because it served as the terminus of the Oregon Trail? From a scientific perspective, the obvious nomination must be the Volcano State, in recognition of the seventeen potentially active eruptive centers that form its geologic backbone—the highest geographic concentration of volcanic landforms in the conterminous United States.

Two of these are well-known: Crater Lake, descendant of the catastrophically born Mount Mazama (see geo-site 67), and Mount Hood, the cauldron of continuing fumarole activity that broods over the city of Portland. A lesser-known geologic site, the Newberry National Volcanic Monument, is home to one of the largest assemblages of cinder cones, lava domes, pumice cones, obsidian flows, and fissure vents in the world. Newberry Volcano may be its crown jewel, but Lava River Cave is its igneous centerpiece—literally speaking. Occasions abound elsewhere for casual strolls across lava landscapes, but here, 12 miles south of Bend off US 97, is an enviable opportunity for a tour into and through the center of a lava flow via the longest continuous, uncollapsed lava tube in the state.

Central Oregon is riddled with at least two hundred lava tubes, roofed conduits of once-molten rock flowing outward from eruptive sites. Lava River Cave, one of the most accessible, meanders for 5,211 feet through the 100-foot-thick, basaltic Bend flow that extends north to the community of Redmond. The Bend erupted 78,000 years ago, from one of the more than four hundred cinder cones and fissure vents that embellish the 500 square miles that encompass the flanks of Newberry Volcano. Detailed mapping indicates its birth vent is situated close to Mokst Butte, about 20 miles southwest of Bend, but the precise locale cannot be determined because it is buried by several younger flows. The flow of molten rock was quickly channeled into several main "streams," in the fashion of flowing water seeking passage across low-lying ground. Eventually, solidifying lava formed a crust over one of these rivers of lava and enclosed the underlying current within a semicircular conduit. When the eruption ceased, the remaining lava drained out of the conduit, leaving behind a lava tube. In time, weathering weakened a section of its roof, which then collapsed, exposing its subsurface wonders to the full light of day.

The dark passages of Lava River Cave feature unusual forms of ornamentation seldom seen elsewhere, such as lustrous, ceramic-like glazed walls formed by incandescent gases; two types of lavacicles—cone-shaped pendants created by lava dripping from the ceiling and delicate, hollow, cylindrical soda straws formed by escaping gas; flow ledges, or shelves, formed when lava accreted to the walls during periods of extended flow; and a rippled floor, a surface created when low-viscosity lava formed a skin that was then dragged into ropy folds by the underlying flow.

This vulcanian subworld is best explored by rented lantern. The self-guided, 40-degree-temperature tour begins at the Collapsed Corridor, proceeds through Echo Hall, Low Bridge Lane, and Two Tube Tunnel, and then terminates in the Sand Garden—a memorable trek along and through what was once a river of molten rock.

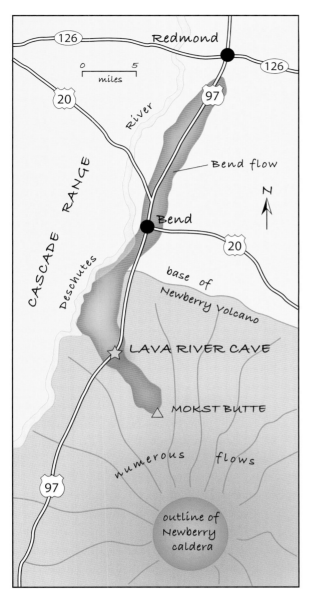

The Bend flow is one of numerous flows that erupted over the past 500,000 years from both the central vent of Newberry Volcano and sites along its flank.

A choked corridor of fallen rock identifies the sunlit entrance to Lava River Cave, which is protected by a roof that averages less than 90 feet in thickness. —Courtesy Lee Schaefer, Sunriver, OR

Flow ledges ornamenting the walls of Lava River Cave are evidence of the currents of molten rock that once passed through this subterranean feature.
—Courtesy of Charles Larson, Vancouver, WA

69. Drake's Folly, Pennsylvania

41° 36' 39" North, 79° 39' 28" West
Devonian Period Petroleum Generation

This first commercial oil well in the United States gave birth to the present-day petroleum industry.

Edwin L. "Colonel" Drake became a man of destiny on Saturday, August 27, 1859, when he supervised the completion of the first borehole ever drilled specifically with the intent of discovering oil. Unknowingly, in the small community of Titusville, he launched the modern hydrocarbon industry. On that occasion America entered day one of its overwhelming reliance on black gold as a source of power, transportation, and myriad other commodities necessary to the modern way of life. Today, after the drilling of several million exploratory wells worldwide, the discovery of more than 2 trillion barrels of crude oil, and the consumption of approximately half that amount, the American way of life is in transition from an era of oil to one balanced between the use of fossil fuels and alternate forms of energy.

The Drake well came perilously close to becoming the folly local residents had snidely predicted. Had the borehole been sited a short distance away, it would not have penetrated oil-bearing sandstone at a depth of 69 feet and 6 inches, because the drill site was very close to the edge of the productive Devonian-age strata. Drake chose the opportune location of the discovery well because it was near a well-known oil seep—one of many that generations of Native Americans had exploited in the region—close to a stream bank. Exploration techniques in those early, simple-approach days focused on the concept of "creekology," the belief that drilling success improved when a well was sited in proximity to flowing water.

Throughout geologic time the formation of petroleum has been generally confined to sedimentary rock deposited in the waters of former oceanic basins in which there existed a rich diversity of microscopic life. When these organisms died, most decayed, but some remains were preserved within seafloor sediment. With the passage of significant periods of time they were buried to depths where the organic compounds were converted by increasing temperatures to hydrocarbons. Eventually, dispersed droplets of crude oil migrated from the organic-rich source rock into an adjacent or shallower reservoir rock comprised typically of permeable sandstone or limestone. The final requirement for forming a pool, or subsurface accumulation, of petroleum is an impermeable caprock to prevent the oil from reaching the surface as an oil seep.

The seminal Drake well produced oil for twenty-six months, at the typical rate of twenty barrels per day. The value of this early production rose to $25 a barrel, equivalent to more than $500 a barrel in 2012 dollars. Soon the surrounding valley was covered with derricks operated by rival wannabe-millionaire oilmen, a number of them producing oil at the rate of several thousand barrels per day. The drop in oil prices caused by this increased production forced Drake out of business, and by 1862 he was unemployed, one of the first victims of oil supply and demand.

A full-scale wooden derrick, replicating the original structure, has been constructed on the actual site of the Drake well, as part of the Drake Well Museum complex at the north end of Oil Creek State Park, on the outskirts of Titusville. The significance of the event that occurred here in 1859 is accentuated by the continuing and positive correlation between world economies and the availability and distribution of immense quantities of petroleum.

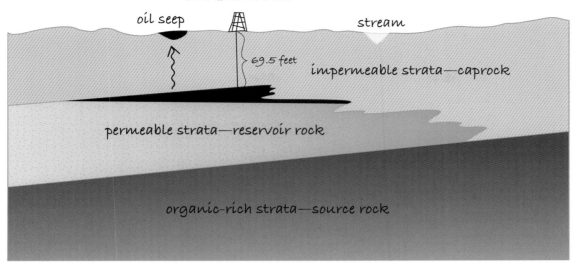

The relationship between source rock, reservoir rock, and caprock at the Drake well site. The oil seep occurred because of the presence of localized fractures in the caprock, which acted as a conduit for a small quantity of oil to migrate to the surface.

This board-for-board replica of Drake's engine house and derrick was constructed using photographs of the actual site of the 1859 discovery well.

70. Hickory Run, Pennsylvania

41° 03′ 02″ North, 75° 38′ 43″ West
Pleistocene Epoch Periglacial Activity

A visual puzzlement, this enigmatic, rock-ribbed carpet of boulders is actually the product of glaciation.

The only consistency in the long, 4,600-million-year history of Earth is change. Mountains rise and wear down. Continents are assembled and then torn by rifting. Climates fluctuate between hot and cold. Today the big news involves global warming, but not long ago in North America global cooling was the story—the great ice age. Pleistocene ice sheets greatly affected the landscape north of the Missouri and Ohio rivers, leaving in their wake a variety of landforms. Less well known are the effects of periglacial environments, those induced by frost action beyond the periphery of a glacier.

A serendipitous combination of mountainous terrain, fractured bedrock, and glaciation molded eastern Pennsylvania into terrain marked by an unusual type of periglacial product: large, barren, jumbled boulder fields. Particularly destitute in appearance, the Hickory Run boulder field, within Hickory Run State Park near the intersection of I-80 and I-476, encompasses within its 18-acre site an estimated 10 million tabular to round boulders up to 30 feet in diameter. Variances in size, shape, and position create an undulating surface that exceeds 4 feet in relief.

At first glance, this rock emporium seems haphazard, but closer inspection reveals a degree of orderliness. Measurements show rock size decreases with depth, while downfield—from east to west—boulder size decreases and boulder roundness increases. Excavations within the field indicate it has a maximum depth of at least 6 feet, and the sound of bubbling water below is at times discernable. The boulder interstices—spaces between individual rocks—are completely free of fine-grained material, a condition that only adds to the field's harsh appearance.

Analysis of aerial photographs indicates this geo-site is composed of a network of linked, elongated, polygonal patterns, each with a core of small boulders. How these sorted-stone rings formed is still disputed, but there is consensus regarding the source of the rocks and their arrangement throughout the boulder field. Their mineral composition and reddish gray color prove they are derived from the Duncannon Member of the Catskill Formation, a Devonian-age sequence of sandstone and conglomerate that underlies and surrounds the area.

The Pleistocene glacier that invaded Pennsylvania 20,000 years ago stopped just short of the Hickory Run boulder field. Temperatures in front of this ice sheet fluctuated daily from above to below freezing, causing freeze-thaw cycles that further cracked the already fissured bedrock into angular boulders. Gradually this field moved downslope by gelifluction, the progressive flow of earth material lubricated by a slurry of ice particles, mud, and sand. In the process, rocks were rounded and became smaller by abrasion, the mechanical grinding away of rock surfaces through friction and impact. Simultaneously, frost-heave cycles lifted large rocks to the surface. Finally, meltwater derived from the retreating glacier cleansed the boulder interstices.

Boulder fields have been studied throughout the world, from Europe where they are termed *felsenmeer*, German for "sea of rocks," to those Charles Darwin famously described in Tierra del Fuego, South America. They form part of the ancient lore of native tribes and for decades were an enigma to scientific interpretation. Geoscientists consider Hickory Run exceptional in comparison to other boulder fields because of its geographic extent, low (1-degree) gradient, diversity of features, and accessibility. It is a genuine rock relic and true leftover from that time when North America was emerging from its most recent big chill.

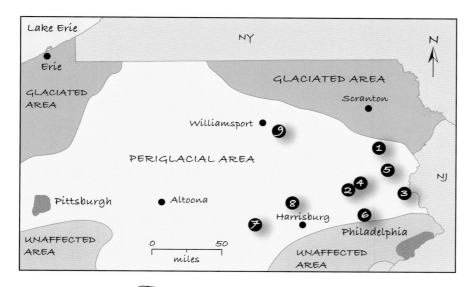

Boulder fields are common in the periglacial terrain of Pennsylvania: (1) Hickory Run, (2) Blue Rocks, (3) Ringing Rocks, (4) River of Rocks, (5) Devils Potato Patch, (6) Ringing Hills, (7) Whiskey Spring, (8) Devils Race Course, and (9) Devils Turnip Patch.

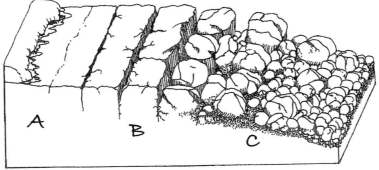

Freeze-thaw conditions altered fractures in the bedrock (A) into blocks (B) that weathered in place to an accumulation of boulders (C). —Courtesy of Pennsylvania State Parks

This geo-site is considered the best example of a boulder field in the eastern United States.

71. Delaware Water Gap, Pennsylvania

40° 58' 13" North, 75° 07' 42" West
Phanerozoic Eon Stream Piracy

Four differing hypotheses are offered to explain this historic, aquatic right-of-way.

The scenery along the drive east on I-80 outside Stroudsburg is breathtaking, but the geology is of a scale ideally viewed from the sky. In a span of 2 miles the highway leaves Pennsylvania, crosses the Delaware River, enters New Jersey, and immediately moves along the tight curve that identifies the Delaware Water Gap, billed by the National Park Service as "one of the best places in America to see the power of water at work."

Found where mountainous terrain and flowing water coexist, a water gap is a deep pass in a mountain ridge through which a stream flows. In contrast, a wind gap is an abandoned water gap forming a shallow notch along a mountain crest. Water gaps and wind gaps are commonly found in geographic proximity to each other. The definitions are uncomplicated, but explanations of how these features formed are not so simple. Multiple working hypotheses exist regarding the origin of gaps, and all have been the topic of extended debate.

The Delaware Water Gap is a 1,200-foot-deep, mile-wide, S-shaped, rounded gorge through Kittatinny Mountain, the easternmost hogback of the Valley and Ridge Province of the Appalachian Mountains. The hard-core spine of Kittatinny consists of the Shawangunk Formation, a flinty, moderately thick sequence of northwest-plunging Silurian-age sandstone and conglomerate. Immediately to the east lies the Great Appalachian Valley, locally called the Wallkill Valley, composed of easily eroded shale and limestone of Cambrian and Ordovician age.

Four different hypotheses attempt to explain how the gap formed:

(1) The Delaware River maintained its course of flow during a period of uplift, sculpting downward seemingly unaffected by the rising topography. This idea implies a river older than the mountain.

(2) Glacial till deposits, such as moraines, blocked major rivers operating in the Stroudsburg area, causing them to pond. Eventually, the rivers broke free of the blockages and eroded across newly exposed topographic divides, forming the dozen gaps that punctuate the 12-mile span between Stroudsburg and Wind Gap, to the southwest. This is the least supported of the hypotheses.

(3) The nature of the rock structure in the area allows—even welcomes—river erosion. On the New Jersey side of the gap, rock strata on Mount Tammany dip at a 50-degree angle to the northwest; however, on the Pennsylvania side the strata on Mount Minsi dip 25 degrees to the northwest. This localized and drastic reduction in rock angle suggests the brittle Shawangunk conglomerate is fractured and jointed in the vicinity of the gap and therefore quite susceptible to erosion.

(4) The hypothesis of choice came into vogue when geoscientists noted the rivers draining the Atlantic side of the Appalachian Mountains had steeper slopes than those draining the midcontinent side. After the tectonic activity that created the Appalachians had ceased, the ancestral Delaware River flowed southeasterly, down the flank and at a right angle to the axis of Kittatinny Mountain. Aided by the high gradient of the Kittatinny hogback and the presence of fractured bedrock, the Delaware gradually extended its course upstream in a process called headwater erosion. The Delaware was able to carve the gap through Kittatinny and then extended its valley even farther upstream until it captured the westerly flow of Cherry Creek.

The ubiquitous distribution of water gaps—more than one thousand are reported worldwide—is lasting evidence of the erosive power of running water. Without them, many vehicular and railroad routes would either have to tunnel through, or deviate around, the mountainous terrains that wrinkle Earth's continents.

River systems commonly develop parallel to the elongated trends of folded rock. At Delaware Water Gap, the Delaware River slices directly across Kittatinny Mountain. This view looks north from the footbridge in downtown Portland, Pennsylvania.

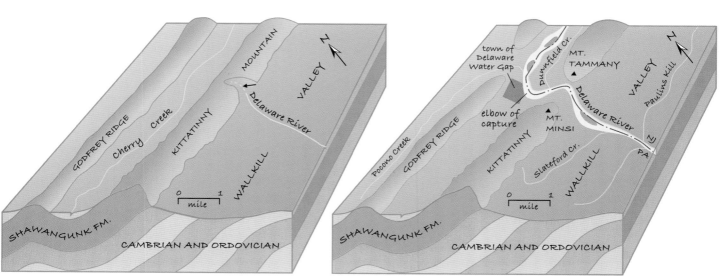

The extension of the valley of the Delaware River by headwater erosion (arrow) was the first stage in the development of the Delaware Water Gap.

Headwater erosion ultimately cut through the entirety of Kittatinny Mountain and captured the flow of Cherry Creek. This process formed a classic elbow of capture, an abrupt or sharp bend where the course of one stream is absorbed into the valley of the capturing stream.

72. Beavertail Point, Rhode Island

41° 26′ 57″ North, 71° 23′ 58″ West
Paleozoic Era Continental Growth

This seaside rock segment is part of one of several exotic terranes that were knitted together into New England.

Guarding the entrance to Narragansett Bay, Aquidneck Island is widely known for the cluster of "summer cottages" that typify the extravagances of the Gilded Age. Here, architecture has captured one era of the nation's human history. Several miles to the west lies smaller Conanicut Island, where a contorted suite of exotic-terrane rocks recall an equally compelling episode in the much longer history of the North American continent. Here, rocks provide an enlightening glimpse into how New England came to be over the 300-million-year span of the Paleozoic era.

Exotic terranes are discrete, fault-bordered rock formations, or groups of related rock formations, that originated elsewhere, were shuffled across the face of Earth by plate tectonic motion, and finally joined a preexisting continent. Since the 1970s, the entire east coast of North America has been remapped as a mélange of accreted exotics, each distinguished by different age, fossil content, and rock type. Outcrops of one of the better known of these miniature landmasses, the Avalon terrane, constitute the rugged shoreline of Beavertail Point at the southern tip of Conanicut Island. The Avalon terrane, also known as Avalonia, began life as a volcanic island arc situated off Gondwanaland, the forebear of today's continents of Africa, South America, and Antarctica. The terrane eventually broke away from Gondwanaland, becoming a drifting microcontinent.

The story begins with the Taconic orogeny, the first of three mountain building episodes that created the modern Appalachian Mountains. During Late Ordovician time a trio of exotic terranes—Piedmont, Dunnage, and Gander—collided with Laurentia, greatly adding to the real estate of this early stage of North America. The second episode, the Acadian orogeny, occurred during late Silurian time, when the Avalon terrane collided with Laurentia and added yet more to the continent's landmass.

The finale of Paleozoic mountain building—the Alleghanian orogeny—reached its crescendo during middle Permian time, when Laurentia and Gondwanaland were tectonically welded together as part of the great supercontinent Pangaea. The Appalachian Mountains, then a collage of contorted and metamorphosed rocks, lay centrally imbedded in this "one Earth" landmass that stretched from the south pole to the north pole. By the time the dust of the Alleghanian collision had settled, folding and faulting had compressed—by as much as 50 percent—the original east-west extent of all four of the terranes that make up New England. The resulting accordion effect created a mountain range that rivaled the height and beauty of today's Himalayas. After Pangaea fragmented during the Mesozoic era, the Appalachian Mountains assumed their position as a prominent coastal range.

The quartet of exotic terranes that parented modern-day New England is still very much in evidence, but only one is represented—and very dramatically so—at Beavertail Point, where the Avalon terrane structure is nakedly exposed, washed by tidal waters. The juxtaposition of light and dark metamorphic rock identifies the fault that slices beneath the old lighthouse foundation, evidence of compressional forces so profound that massive degrees of fracture were the result. The U-shaped fold 50 feet southeast of the lighthouse foundation illustrates forces as significant, but less intense in application. These emigrant exposures of rock have experienced much history since their Precambrian birth thousands of miles to the south. The long-term survival of these unyielding metamorphosed rocks is testimony to their resistance to erosion.

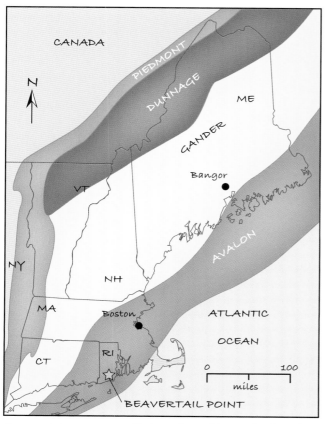

Four exotic terranes are welded together as New England: Piedmont, Dunnage, Gander, and Avalon. Avalon rock underlies all of Rhode Island. (Modified after Raymo and Raymo, 2001.)

The Beavertail Fault (dashed line) divides black-hued phyllite (a metamorphic rock intermediate in grade between slate and schist) of the Dutch Island Harbor Formation (on the right) from light green to buff-colored phyllite of the Jamestown Formation.

This fold at Beavertail Point is representative of the landscape of contorted and disturbed rock that characterizes the Avalon terrane.

73. Crowburg Basin, South Carolina

34° 46′ 58″ North, 80° 24′ 32″ West
Triassic Period Continental Rifting

The southernmost exposure of a long line of rift basins that witnessed the birth of the Atlantic Ocean.

From a geological perspective, any drive along the eastern seaboard of the United States is an ideal opportunity to study the strata involved in the birth of the Appalachian Mountains. These forest-covered slopes are constructed principally of sedimentary rock, tens of thousands of feet of it deposited under fluctuating oceanic conditions that lasted for almost 300 million years. This is a land of limestone, coal, sandstone, and shale; a palette of gray, green, taupe, and black; and a story of rock folding and crustal compression, faulting, and shortening. Here generations of students have conducted chapter and verse analysis of an important segment of Earth history, beginning with the introduction of skeletal life and ending with the amalgamation of the supercontinent Pangaea.

To the armchair geologist this is "Paleozoic land," pure and simple. A more thorough investigator discovers a younger chapter written in terms of basins filled with brick red conglomerate, maroon shale, and rust red siltstone. These rock exposures, oriented northeast to southwest along the eastern seaboard, intermittently indent the eastern approaches to the Appalachians from Nova Scotia to South Carolina. Farther south, a far as Florida, similar basins are buried beneath younger rocks forming the Atlantic Coastal Plain. The basins are the land of post-Paleozoic, Triassic-age red beds, which represent a story of fracture and crustal extension and lengthening—a story that contrasts with that of the building of the Appalachians.

Vast chambers of liquid rock rising from the core of the Earth accumulate within the roots of continents and supercontinents to simmer and boil, first doming, then stretching, and finally fracturing the overlying crust. As the continent stretches, the fractures widen into fault-edged rift basins that first fill with nonmarine sediments, then become even wider and deeper, allowing the eventual invasion of saltwater seas. The deposition of marine sediments signals a new ocean basin has been born. In a nutshell, this is the plate tectonics process, the discovery of which revolutionized geology in the 1960s.

The Triassic red-bed basins that accessorize the Appalachian slopes represent the nonmarine phase in the birth of the Atlantic Ocean. Geologists have identified almost two dozen of these 235-million-year-old troughs trending subparallel to the Atlantic Coast. Many are exposed, while the rest lie under a blanket of marine rocks deposited when unusually high seas invaded the region during the Cretaceous period.

One of the exposed basins garners special attention. Compared to its much larger northern kin, such as the Newark Basin of New Jersey (see geo-site 53) and the Hartford Basin of Connecticut (see geo-site 15), the Crowburg Basin, covering 7 square miles in Lancaster County, is minuscule and visually disguised. However, three accessible exposures readily define this red-bed evidence of continental extension and growth:

(1) 0.9 mile northwest of Pageland beside South Carolina 207, the contact between volcanic rock and maroon and red conglomerate and sandstone in an outcrop identifies the fault—the Pageland Fault—forming the southeastern edge of the Crowburg Basin.

(2) 1.9 miles farther northwest is an exposure of sedimentary rocks typical of the nonmarine strata that fill this basin.

(3) 0.1 mile north of the intersection of South Carolina 9 with road 13-545 (6 miles west of Pageland), opposite Henderson Road, a second exposure of the Pageland Fault contact further defines the basin border. The red beds exposed here are lauded as the southernmost outcropping of Triassic sedimentary rock in the eastern United States.

To the field geologist the Crowburg Basin offers convincing evidence of how a small and unassuming countryside exposure can be large and of great importance in the interpretation of geologic history.

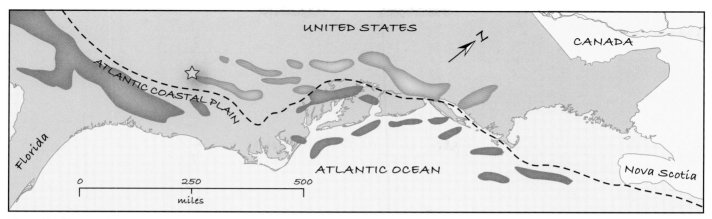

The red-bed basins of Triassic age are divided among those that are exposed (orange) and those buried beneath the Atlantic Coastal Plain (gray). The Crowburg Basin (star) is the southernmost of the red-bed basins that are exposed.

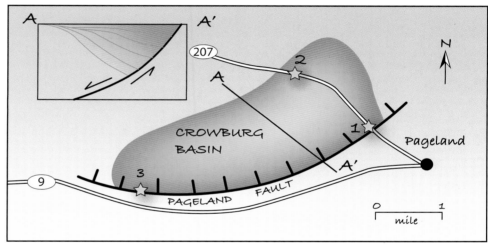

Map and cross section of the Crowburg Basin showing the location of three identifying outcrops and the trace of the basin's southern boundary delineated by the Pageland Fault. The barbs on the fault occur on the fault block that dropped relative to the fault block on the other side; arrows in the cross section denote relative direction of movement along the fault.

Close-up of the red-bed conglomerate, shale, and sandstone that fill the Crowburg Basin, exposed north of Pageland at site 2. The stake is 15 inches high.

165

74. Mount Rushmore, South Dakota

43° 52′ 45″ North, 103° 27′ 35″ West
Proterozoic Eon Magma Intrusion

Four fabled presidents gaze eternally outward from the face of 1.7-billion-year-old fractured granite.

While a relative lack of relief characterizes South Dakota, the southwest quarter is an exception. Here, the Black Hills punctuate the Great Plains as a tectonic preamble to the Rocky Mountains. The geology of this topographic outpost is rather simple: sedimentary rock of Paleozoic and Mesozoic age encapsulates an oblong metamorphic halo, which in turn surrounds an igneous core. In this scenic heart, between the cities of Keystone and Custer, lies 7,242-foot-high Harney Peak, the loftiest point in the state, and Mount Rushmore National Memorial, the state's number one tourist attraction. Here, geologic history, a paean to American leadership, and the art of sculpture are beautifully intertwined.

A shallow sea once blanketed this land with quantities of sand, mud, and clay that were later compressed to sandstone and shale. Then, some 1,700 million years ago, a monster-mass of molten rock intruded the Precambrian scene, radiating heat that transformed the sedimentary rock to its metamorphic equivalents of quartzite and schist. Fingers of molten rock opportunistically probed avenues of fracture. All the while, the magma soup consumed chunks of preexisting rock. As this maelstrom cooled, late-phase injections of magma filled crevices—their numbers are legion—to form dikes. When crystallization had run its course, the deeply buried bulk of Harney Peak granite lay confined within a metamorphic aureole.

During the ensuing billion years, erosion was the dominant force, removing some of the rock above the granite mass. With the dawn of the Paleozoic era, however, seas again began to ebb and flow across the land, depositing a 7,500-foot-thick pile of sedimentary rock. Finally, 70 million years ago the entire region began to rise in concert with the birth of the Rocky Mountains, and erosion removed the sedimentary rock veneer, exposing the crystalline core—the Harney Peak granite and surrounding metamorphic rock. The Black Hills emerged. Today, frost wedging, a process in which water repeatedly freezes and thaws in the fractured rock, enlarging the fractures, continues to modify the profile of this topographic blister.

The concept of portraying historic figures on the heights of the Black Hills dates back to 1923. Once the site of Mount Rushmore and a great presidents theme were chosen, a three-stage preparatory process began. Workers first used dynamite to remove 400,000 tons of deeply weathered Harney Peak granite. Then, with hammer and chisel, they carved away an additional 3 to 6 feet of rock to prepare for the final step, during which pneumatic drills "bumped" the granite surface to the smoothness of sidewalk concrete. With a several-hundred-foot-thick column of prepared granite now exposed atop a shelf of older schist, the artistic phase could begin. In 1941, Mount Rushmore was prematurely declared completed—funding had run out due to the onset of World War II.

Multiple sets of light-colored dikes "age" each presidential face, Thomas Jefferson the most. The face of Theodore Roosevelt is greatly inset because 80 feet of severely cracked surface rock had to be removed from the area between his ears. Each year, silicone caulk is used to fill any microfractures that develop, and in 2005 workers washed the faces for the very first time. At the rate this memorial is deteriorating, estimated at 1 inch every 10,000 years, it will surely be many years, and many generations, before these 60-foot-tall countenances require a cosmetic facelift.

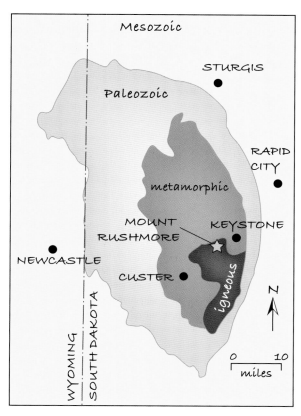

Geologic map of the Black Hills. Sedimentary strata of Paleozoic and Mesozoic age surround a core of igneous and metamorphic rock.

Coarse-grained dikes that intrude the finer-grained Harney Peak granite mar the face of Thomas Jefferson. —Courtesy of Galen R. Frysinger, Sheboygan, WI

The bulk of Mount Rushmore is composed of Harney Peak granite that intruded older layers of metamorphic rock (highlighted). A talus slope of waste rock (highlighted) fills the foreground.

75. Mammoth Site, South Dakota

43° 25′ 29″ North, 103° 28′ 58″ West
Pleistocene Epoch Fossilization

A 26,000-year-old sinkhole preserves the linebacker megafauna of the most recent ice age.

Nestled among the rolling hills of southwestern South Dakota and named for its thermal waters, Hot Springs is home to an eye-opening, back-to-the-past window into the most recent ice age: the Mammoth Site. Located at 1800 US 18 Bypass, this 120-by-150-foot, 65-foot deep, sand- and clay-filled sinkhole is acclaimed as the largest concentration of mammoth remains in the world.

Mammoths were large, elephant-like mammals that evolved in Africa, migrated to Europe and Asia, and crossed the Bering Land Bridge that connected eastern Siberia to North America. At the height of their reign, they grazed and browsed as far south as Central America. They became all but extinct 10,000 years ago, possibly due to climate change and overhunting by the Paleo-Indian Clovis culture. Paleontologists have discovered evidence that remnant herds populated a small Arctic Ocean island as recently as 4,000 years ago.

Two early versions of the proboscidean—characterized by a trunk—family proliferated on the American prairie: the Columbian and its smaller first cousin, the woolly. The Columbian was a brute that stood 13 feet tall—so large the contemporary African elephant could walk under its chin—weighed in at 10 tons, and spent most of its life seeking food and drink. Small in comparison, the woolly topped out at 10 feet and 6 tons. Both were vegetarians; fossilized dung analyses indicate they preferred grass and sedge.

Researchers used a tusk count of two per animal to identify the remains of fifty-two male Columbian and three male woolly mammoths, the majority of them young, at the Mammoth Site. One mammoth, dubbed "Napoleon," is intact bonewise, but the headless "Murray" may be the star of the exhibit. Murray was originally named Marie Antoinette, the moniker having been changed upon discovery that she was, in fact, a he. These mammal remains are not fossils in the normal sense of having been permineralized, a process whereby mineral matter replaces the original organic components. Instead, they are composed of original bone and therefore quite fragile, so many have been left on exhibit exactly as they have lain since the moment of death. Luckily, neither scavengers nor river currents have disarticulated the skeletal remains.

Southwestern Hot Springs is built on a bedrock foundation composed largely of late Paleozoic limestone that is very susceptible to being dissolved by groundwater. Approximately 26,000 years ago the roof of a cavern that had developed in this rock due to dissolution by artesian springwater collapsed. Over the course of 700 years, as the steeply inclined sinkhole slowly filled with sediment, this bubbling artesian spring attracted countless mammoths, along with numerous other ice age mammals. For the unwary, a site valued for life-sustaining water unfortunately was a site where loss of life by entrapment in gooey mud became a final chapter—long-term incarceration in a mammoth mausoleum.

Today the Mammoth Site is housed under the roof of a major research repository, where excavations continue on a seasonal schedule, unleashing evidence of a paleoenvironment where the arctic woolly lies entombed with the temperate Columbian. This fact alone justifies the curator's claim that this is not merely a great museum, it is a *mammoth* museum.

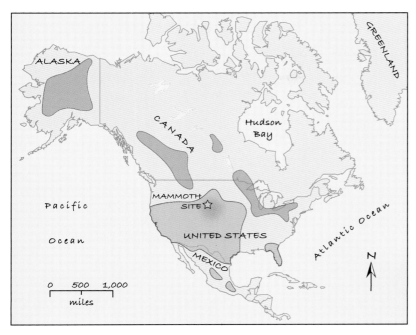

Generally, Columbian mammoth fossils are found in the United States, Mexico, and Central America (orange), while woolly mammoth fossils are principally found in the Great Lakes region north to Alaska (green).

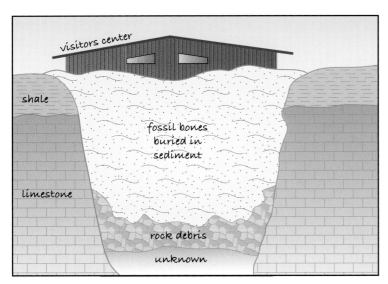

Rock debris derived from adjacent beds of limestone and shale fill the depths of the Mammoth Site sinkhole. Additional sediment buried and preserved masses of fossil bones.

The front limbs of "Murray" reach upward in an unsuccessful attempt to free himself from his sinkhole tomb. —Courtesy of Joseph Muller, Mammoth Site, SD

76. Pinnacles Overlook, South Dakota

43° 52′ 10″ North, 102° 14′ 06″ West
Quaternary Period Wind and Water Erosion

Powerful terrestrial erosion changed the good lands of the western plains to badlands.

There is a tremendous amount of geology in the world—seven continents' worth, plus more beneath the oceans. Libraries and Web sites overflow with reports on the subject, all too often complicated or even incomprehensible. Fortunately, there are areas where the beauty and orderliness of the natural world are portrayed in uncomplicated transparency. Pinnacles Overlook, at the western end of Badlands Loop Road in Badlands National Park, is one such locale. Here, the geologic saga of the South Dakota plains can be decoded in terms clear and concise because the entire sequence of horizontally layered rocks that make up the Badlands is open to view.

The heart of the Badlands is a sequence of flat-lying sedimentary rock that is both a model of desolation and an orchestrated kaleidoscope of color. Divided into three major units, the Chadron, Brule, and Sharps formations, this sequence began its history 33 million years ago, when the Western Interior Seaway had loosened its grip on North America and dinosaurs had departed the scene.

Initially, a subtropical climate bathed the land, which lay beneath a thick, canopied forest, while rivers deposited layers of mud and sand across the uneven terrain. Alligators, tortoises, and rhinoceros-like herbivores roamed in uneasy association. In the immutable geologic cycle, sediment became rock—the Chadron Formation. Composed principally of greenish clay, the Chadron erodes to gentle, haystack-shaped slopes. When wet the slopes are quite slick, and when dry the clay shrinks to particles that look like popcorn.

When the buff-colored siltstone and sandstone layers of the Brule Formation were being deposited, the climate had become cool and somewhat arid. Clouds of ash blew across the landscape, and rivers of mud entrapped large numbers of the resident life. Recognized today as perhaps the richest repository of Oligocene-age mammal fossils in the world, these beds harbor more than 150 genera, from diminutive rodent to hefty rhinoceros. This was the time when savannah replaced forest and the cud-chewing oreodont, a sheeplike grazer whose population perhaps numbered in the millions, occupied center stage. Brule strata are easily recognized because they erode to steep, knifelike ridges, pinnacles, and overhangs that form a fairy castle topography accentuated with purple, gray, and crimson bands, which can appear almost translucent after a summer thunderstorm.

The youngest unit of the Badlands pile of rock is the Sharps Formation, characterized by layers of dazzling white volcanic ash dating to approximately 27 million years ago. This wind-borne material, best viewed at the Pinnacles Overlook, marks the end of the depositional phase of the Badlands story, by which time hundreds of feet of new rock had been added to the "good land" surface of the western plains. The infamous serrated topography of today, however, did not then exist. Some 500,000 years ago uplift rejuvenated the High Plains, giving its rivers the energy to seriously sculpt the terrain. This was the beginning of the badland phase, which is characterized by episodes of wind and water erosion that continue today.

The savage rock configurations of the Badlands weather away, on average, 1 inch per year. Should this pace continue, they will be gone within another 500,000 years. The Badlands story can then be summarized as 10 million years of deposition followed by 1 million years of erosion. In the long history of Earth, this cycle is repeated time and again. In nature as in life, it takes much longer to build than to destroy.

Pods of limestone scattered throughout the layers of multi-colored strata of the Chadron Formation (foreground) are evidence that lakes freshened the landscape of Eocene time.
—Courtesy of Ed Faires, Kernersville, NC

At Pinnacles Overlook the entire Black Hills column of rock is on view: the Chadron deep in the valleys, the Brule centrally located, and the gleaming white ash beds of the Sharps Formation in the foreground.

Purple and crimson bands, variable bed thickness, and knife-edged strata are distinguishing features of the windblown sand, water-deposited clay, and volcanic ash that characterize the Brule Formation.

77. Reelfoot Scarp, Tennessee

36° 26' 14" North, 89° 26' 22" West
Holocene Epoch Earthquake Movement

Two centuries ago, a near-record displacement caused Old Man River to temporarily flow upstream.

More than 3 million earthquakes occur each year—on average one every ten seconds. Eighteen qualify as major, registering a magnitude (measure of released energy) of 7 or more. One is catastrophic (magnitude 8). In recent U.S. history, three stand out: 1886 in Charleston, South Carolina (magnitude 7); 1906 in San Francisco (magnitude 7.8); and 1994 in Los Angeles (magnitude 6.7).

These occurrences pale in comparison to the seismic events that unfolded December 16, 1811, changing forever the landscape of the frontier river town of New Madrid. Over the following eight weeks more than two thousand temblors rumbled throughout the New Madrid Seismic Zone (NMSZ) and were felt as far away as Boston, Massachusetts. Surviving accounts record a heaving and cracking Earth, swamped and sinking boats, the instantaneous rupture of massive tree trunks, and a rain of sandy debris exploding from surface craters. Several river islands in the Mississippi abruptly disappeared, and two small frontier towns were completely destroyed.

The culminating act of this two-month reign of destruction was worse. At 4:00 a.m., February 7, 1812, a growing roar heralded ground fissuring to a depth of about 12 feet, extensive landsliding, water fountains spouting 8 to 10 feet, trees snapping and falling by the thousands, and sulfurous odors contaminating the air. On the Tennessee side of the Mississippi River the Reelfoot Fault, one of many that compose the NMSZ, hatched at a depth of 8 miles and tore the surface, creating a 20-mile-long fault escarpment. When the ground on the east side of the fault collapsed, a local basin was created that eventually filled with water, forming 25,000-acre Reelfoot Lake. Simultaneously, the fault movement crossed the course of the Mississippi River and reversed its gradient for a distance of some 10 miles, causing it to do the seemingly impossible for several hours: flow to the north—upstream. Quickly, however, the power of flowing water overcame the retrograde motion of the river.

Modern analyses of this destruction, which occurred prior to the existence of scientific instrumentation, suggest that during the February 7 quake magnitudes exceeded an almost unbelievable 8.5. Many geologists consider these earthquakes of 1811–1812 the most intense in the history of the contiguous United States, and the NMSZ the most dangerous earthquake region in the country. For decades, no one knew why the quakes were centered in the North American continent, rather than along the Pacific Coast where such activity is historically concentrated. Answers are now available.

The Reelfoot Fault dates to 750 million years ago, when extensional fracturing of the supercontinent Rodinia created the NMSZ, a regional system of faults associated with the opening of the Iapetus Ocean. About 500 million years later the breakup of another supercontinent, Pangaea, gave the NMSZ a second life. With North America, born from the breakup, today drifting slowly westward, the presence of deeply buried, decaying but not necessarily dead internal forces give rise to periodic tectonic activity beneath the New Madrid region, creating an odd phenomenon: a seismic zone located midcontinent.

Although some two hundred low-magnitude earthquakes jar the NMSZ annually, the region is relatively quiet, and the once-impressive Reelfoot Fault scarp has been muted by two centuries of erosion by wind and water. The gentle 20-foot rise rambling across a farm field north of Tiptonville, however, is evidence of rock and roll forces unleashed that long-ago night when geologic history was made. The scarp is visible from Van Works Road, which is roughly 5.5 miles north of Tiptonville off Tennessee 78. Can it happen again? Research shows similar violent quaking occurred in the NMSZ about AD 900 and again around AD 1350. If this 450-year cycle is repeated, Old Man River might again flow uphill around the year 2260. Should this happen, the geologic adage "The present is a key to the past" might well be reworded to "The past is the key to the future."

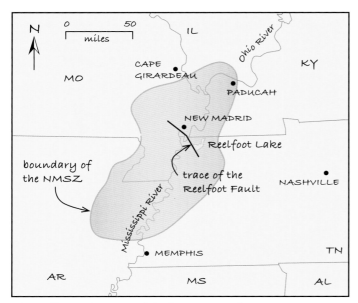

The Reelfoot Fault, origin of the destructive earthquakes of 1811–1812, is one of many faults that lie within the general boundary of the NMSZ. (Modified after Stewart and Knox, 1996.)

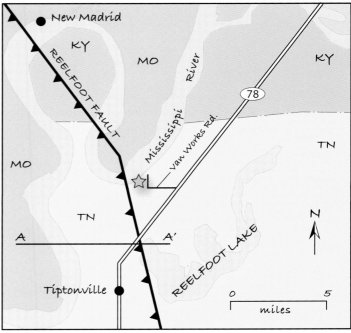

From a position along Van Works Road north of Tiptonville (star) the weathered profile of the Reelfoot Fault scarp is visible on the near horizon to the west. The triangular barbs identify the uplifted fault block.

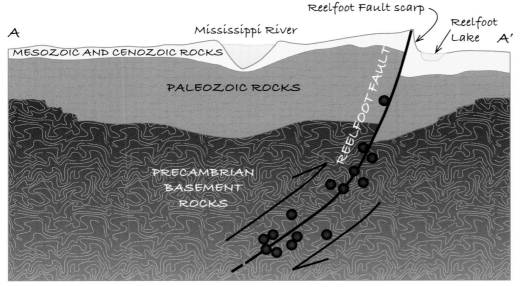

Cross section showing the Reelfoot Fault and the relationship to its scarp and Reelfoot Lake. Arrows denote direction of movement. Even today earthquakes (dots) continue to jar this region. See map above (right) for location of cross section. (Modified after Johnston and Schweig, 1996.)

78. Enchanted Rock, Texas

30° 30′ 22″ North, 98° 49′ 05″ West
Proterozoic Eon Magma Intrusion

Exfoliation created and maintains the rounded profile of this Precambrian bull's-eye of Texas.

Between Houston and Austin the roadside appearance of southeast Texas undergoes a transformation. The uninspiring terrain of the Gulf Coastal Plain gradually changes to the undulating relief of the Hill Country. Limestone and dolomite dominate the basement rock with but one exception: the late Precambrian igneous and metamorphic landscape of the Llano Uplift. This topographic prominence, composed of the oldest rocks found within the Lone Star State, is cartographically unusual because its outcrop pattern forms a bull's-eye on the geologic map of Texas.

More than 1 billion years ago, the region that is now Central Texas was part of an ocean that functioned as a collecting basin for sediments eroding from nearby highlands. Slowly, this area became intimately involved in the global assembly of landmasses that created the earliest well-known supercontinent, Rodinia. Mountains rose to lofty heights, and deep-seated forces metamorphosed some subsurface rock into gneiss and schist, while the rest evolved into massive chambers of liquid rock that percolated upward and crystallized to granite.

As the Paleozoic era dawned and Rodinia began to break up, a new ocean covered Texas. Within it sediment varying in age from Cambrian to Carboniferous was deposited, burying the now-aging Precambrian mountains. Then, in yet another episode of continental assembly and dissection, a theme that threads through the majority of geologic history, the most recent of the supercontinents—Pangaea—came into existence. As this "one Earth" mass underwent its inevitable tectonic breakup, a youthful North America was separated from a developing Europe, the Gulf of Mexico was born, and Texas was inundated for the last time. In this sea a thick blanket of Cretaceous-age Edwards limestone covered the erosion-flattened terrain. Finally, one last tectonic thrust uplifted the ancient Precambrian basement several thousand feet, giving birth to the Edwards Plateau. Erosion over the past 12 million years has dissected this elevated landscape and in the process exposed its ancient core of igneous and metamorphic rock. This exhumed massif, known as the Llano Uplift, is the sole window that allows us access to the Lone Star State's igneous and metamorphic basement.

The polished crystalline hills of Enchanted Rock State Natural Area, 24 miles south of Llano and adjacent to Ranch Road 965, are an ideal location to study both the Precambrian rocks and the granitic terrain that compose the Llano Uplift. Within this park, the 1,825-foot-high Enchanted Rock is but a very small segment of the much bigger Enchanted Rock stock, which together with six similar stocks comprises the core geology of Central Texas. The granite of Enchanted Rock is 1,048 million years old and is locally prized as an architectural stone, rock used to beautify the exterior of buildings.

The flanks of Enchanted Rock are littered with textbook examples of concentric slabs of rock formed by a weathering process called mechanical exfoliation (see geo-site 18). As erosion gradually exhumed this body of granite, removing overburden and reducing the pressure on the granite, this mass expanded and split into curved sheets that slowly fractured and weathered away, creating the rounded profile of Enchanted Rock. Using the Summit Trail, it is easy to climb Enchanted Rock in one hour. The horizon-to-horizon view from the top is illustrative of how igneous and tectonic activity can greatly reconstruct and enhance otherwise uninspiring topography.

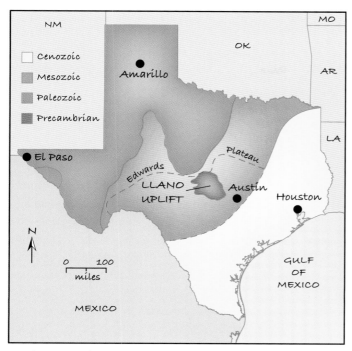

Geologic map of Texas.

Enchanted Rock fits the definition of an inselberg, a prominent, isolated, and rounded knob of rock surrounded by a lowland erosion surface.

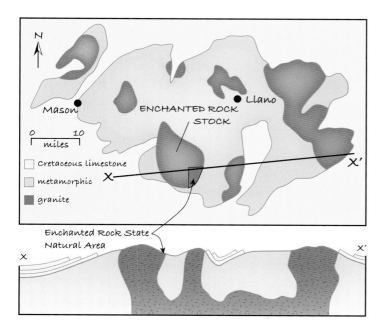

Geologic map of the Enchanted Rock region and cross section showing how erosion has locally removed the original cover of Cretaceous-age limestone. (Modified after Barnes, 1988.)

Exfoliated slabs of granite peel away from the surface of Enchanted Rock, similar to the way an onion separates layer by layer.

175

79. Capitan Reef, Texas

31° 53' 39" North, 104° 49' 20" West
Phanerozoic Eon Reef Development

West Texas hosts a world-class chain of tropical sea reefs containing a treasure trove of black gold.

The latest news articulating the biologic status of living reefs does not bode well for these colonial denizens of the world's oceans. Over the past four centuries coral growth throughout the Great Barrier Reef of Australia—the largest in the world—has declined to a rate described as both unprecedented and sluggish, the result of global warming and increasing water acidity. This is a sad commentary, because reefs are dominant players in the creation of marine ecosystems that serve as central food sources for literally thousands of symbiotic organisms.

Reefs have existed for 500 million years (see geo-site 24). Numerous fossil examples provide evidence that clear, oxygenated, shallow, tropical waters have long graced the seascapes of Earth. An illustrative example is the 400-mile-long Capitan Reef that surrounds the Delaware Basin, an oil-laden province of sedimentary rock deposited across the border between Texas and New Mexico.

The view from the 5-mile-long S-curve of US 62/180, south of the entrance to Guadalupe Mountains National Park, constitutes a panoramic collage of one of the best examples of an ancient fossil reef on Earth. The exhumed brow of Capitan Reef majestically looms over the northern limits of the great Chihuahuan Desert—truly a time capsule exposure of rock that contrasts the blue-water environment that once existed with the near-barren, arid conditions that prevail today.

More than 250 million years ago, the supercontinent Pangaea straddled the equator. Its western shoreline was indented by a series of shallow-water embayments, finger-like extensions of the all-encompassing Panthalassa Ocean. In one of these, the Delaware Basin, the combination of water depth, temperature, and chemistry was just right for the formation of a reef. Unlike the coral that constitutes modern-day reefs, the principal architects of this Permian-age representative were sponges, encrusting algae, bryozoans, and bivalve clams. As the Delaware Basin gradually sank in response to compressional forces associated with the formation of the ancestral Rocky Mountains, this formidable living wall of marine organisms built upward, ultimately reaching a thickness of 2,200 feet.

Eventually, an influx of sediment smothered the Capitan Reef, and the evaporation of its host ocean essentially choked it to death. The huge organic halo that had once identified the shoreline lay buried within a thick insulation of sediment. Finally, some 50 million years ago this region was uplifted, reincarnating the forces of erosion. Today, the most extensive outcrops of the reef lie within the boundaries of Guadalupe Mountains National Park. Smaller exposures, such as the Apache and Glass mountains, exist to the southeast.

Two contrasting vantage points within the park offer an optimal perspective on geologic details of this geo-site. The first is from the 8,749-foot-high crest of Guadalupe Peak, the highest point in Texas. The other is along the McKittrick Canyon Trail to the historic Pratt Cabin, where three distinct reef environments are grandiosely exposed in the walls of the canyon. Back-reef strata deposited in tidal flats and outcrops of broken reef rock deposited in the high-energy fore reef envelop massive exposures of reef-front limestone.

Today, the rock foundation of Capitan Reef is known as both a storehouse of energy—almost 2 billion barrels of oil have been extracted to date—and the most scientifically analyzed example of how fossil reefs define the transition from land to sea.

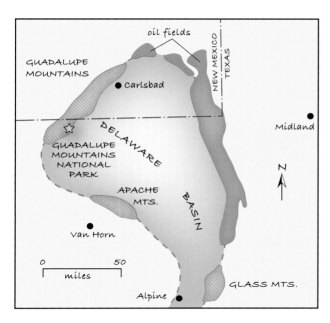

The surface (hachured areas) and subsurface (dashed orange) trend of Capitan Reef define the boundaries of the Delaware Basin. Oil fields further define the basin.

The sandstone slopes and the cliff-forming limestone wall of El Capitan (left) and its sibling prominence Guadalupe Peak (center) accentuate Capitan Reef.
—Courtesy of Michael Haynie, National Park Service

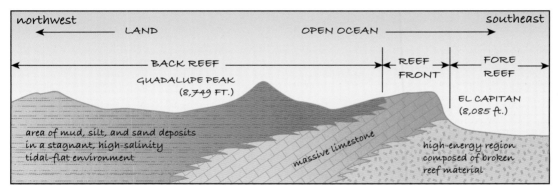

Cross section showing the types of rock found in the three different environmental zones that existed during the growth of Capitan Reef.

80. Paluxy River Tracks, Texas

32° 15' 12" North, 97° 49' 05" West
Cretaceous Period Fossilization

Expansive herds of both hunter and hunted left their tracks along the shoreline of an ancient sea.

One of the most ballyhooed recent news reports involves the increase in atmospheric carbon dioxide and its effect on global warming. The scientific community estimates the average surface temperature of Earth has risen 1 degree Fahrenheit over the past century. A further 3- to 10-degree increase is possible by the year 2100. Add the fact that the ten warmest years on record have all occurred within the past two decades, and these data take on an aura of unprecedented change. Global sea level has risen as much as 6 inches over the past century, the result of melting glaciers. Should this trend continue until all the world's ice is reduced to liquid, sea level will rise approximately 240 feet, reaching the armpits of the Statue of Liberty and flooding 15 to 20 percent of Earth's land surface.

These statements may be troubling, but to geologists they are yesterday's fact. After all, the short form of Earth history reads simply: "The seas came in and the seas went out." A fascinating example of this continuum of eustatic events—worldwide sea level changes—lies in the fossil-bearing rocks of the Glen Rose Formation, exposed throughout north-central Texas. The depositional environments of these limestone strata ranged from nearshore to deepwater sectors of a sea that periodically inundated the land during Cretaceous time.

In Dinosaur Valley State Park, along the banks of the Paluxy River and southwest of Fort Worth between Stephenville and Glen Rose, exposures of Glen Rose strata tell a story that intimately links sea level fluctuation with 110-million-year-old footprints representing three major branches of the dinosaur family. In their heyday, these lizards roamed freely through north-central Texas, kings and queens of their domain, leaving footprints in the viscous calcium carbonate paste of the shoreline. Subsequent burial condensed this mushy mixture into limestone and preserved the tracks as fossils, which today are being unearthed by the waters of the Paluxy River.

Serpentine-necked herbivores—plant eaters—made the largest of these tracks. Their four-toed hind feet left saucerlike, 3-foot-long imprints that closely resemble those of modern-day bears, whereas their smaller, front indentations are like those of horse hooves. These are believed to be the fossil tracks of a *Paluxysaurus*, a 60-foot nose to tail, 20-ton relative of the familiar *Brachiosaurus*. In contrast, the 30-foot long, 3-ton, carnivorous *Acrocanthosaurus*, a relative of the infamous *Tyrannosaurus rex*, left 20-inch-long imprints with three claw-tipped toes. These meat eaters were constantly seeking prey and used their jaws full of serrated teeth to kill and disembowel their victims. A two-legged plant eater belonging to the *Iguanodon* genus probably made the rarely seen, stubby, three-toed prints.

The carnivores of Dinosaur Valley State Park came to the area principally to feast on shore-dwelling animals, whereas the herbivores—because their tracks occur side by side—may have been moving through as a herd on a migratory path. One chilling drama of Cretaceous life is visible in a particular set of fossil trackways. Analysis of stride suggests a scenario in which an *Acrocanthosaurus*, running at 5 miles per hour, was bearing down on its ponderous prey *Paluxysaurus*, capable of only half the speed. The outcome was inevitable—death followed by dinner.

Geology of north-central Texas during Cretaceous time, when calcium carbonate precipitated from a sea and accumulated as a pasty deposit. The tracks at Dinosaur Valley State Park were left as dinosaurs walked through this pasty material across a shoreline that fluctuated in response to changes in sea level.

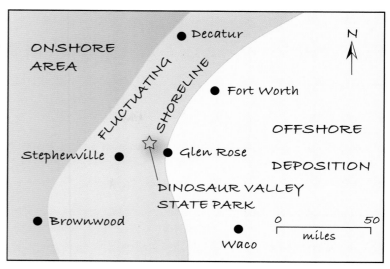

The left hind-foot track (circled in foreground) and crescent-shaped forefoot track (circled in background) of Paluxysaurus. *Three-foot tape measure for scale.* —Courtesy of Dr. James Farlow, Indiana-Purdue University, Fort Wayne, IN

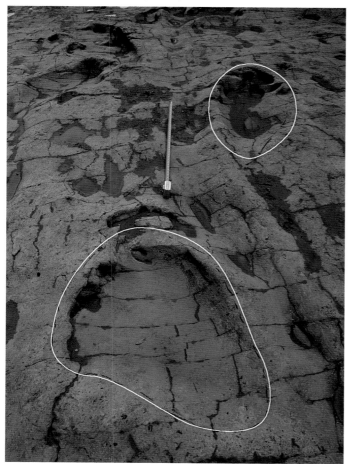

Imprints of the two-legged Acrocanthosaurus *indicate they walked in a pigeon-toed fashion, while stride length suggests they could run at predator-like speeds of more than 10 miles per hour. Yardstick for scale.* —Courtesy of Kathy Lenz, Dinosaur Valley State Park, TX

81. Upheaval Dome, Utah

38° 26' 13" North, 109° 55' 46" West
Paleogene Period Meteor Impact

Controversy has long surrounded this best example of a complex meteor crater known on Earth.

The Colorado Plateau is often called the most spectacular physiographic province in North America, where striking features intermingle with a layer-cake sequence of strata. Nowhere is it more visually compelling than in the eastern portion of Utah. To the west lies the disturbed Basin and Range Province. To the north, south, and east, the Rockies form the central segment of the Backbone of the Americas, those mountains that extend from Alaska to Patagonia.

Amid this barren, red rock world of butte, meander, esplanade (a level, open stretch of ground), and mesa, the circular structure of Upheaval Dome forms the northern cornerstone of Canyonlands National Park, southwest of Moab. It looks like a gigantic bull's-eye, 3 miles in diameter, consisting of a huge, 1,100-foot-deep circular pit surrounded by uplifted rock. Earth-toned shale and crossbedded sandstone of early Mesozoic age that surround the pit dip away from it at angles greater than 30 degrees to create a curvilinear ridge-and-valley structure, which, from an aerial perspective, appears as a fractured geologic blemish. At its core, near-vertical White Rim sandstone of Permian age appears in stark contrast to the thick sequences of horizontally stacked sedimentary rocks that form the surrounding Colorado Plateau.

Early geologists, adapting a hypothesis then in vogue, suggested Upheaval Dome formed when gaseous magma broached the surface as a cryptovolcanic—concealed volcanic—explosion, resulting in a circular pattern of rocks blasted upward and outward. Subsequent field research, however, quickly undermined this scenario.

An alternate hypothesis involved nearby Moab Valley, 20 miles northeast of Upheaval Dome, where a landlocked sea had evaporated during Pennsylvanian time, leaving thick sequences of gypsum and halite (salts) of the Paradox Formation. Subsequently buried by deposits of shale and sandstone, these precipitate minerals began to flow upward because of their lower density. As they came into contact with near-surface groundwater they dissolved, creating a subsurface void. The collapse of overlying sedimentary rocks into this void is believed to have formed the elongate, northwest-southeast-oriented Moab Valley. Since gypsum and halite also underlie Upheaval Dome, could it be a result of the same salt movement? Probably not, because the surface morphology is wrong: the linear geometry of Moab Valley is in direct contrast to the circular nature of Upheaval Dome.

By the 1960s the focus had turned to the possibility of an extraterrestrial origin. The circular arena of highly disturbed rock could be evidence of meteor impact, but additional proof was needed. Field mapping eventually identified a field of shatter cones, a type of disturbed rock, near the core of the dome. These grooved, cone-shaped features are created when rock is subjected to stresses up to 200,000 times greater than atmospheric pressure—stresses attributed to meteor impact. Recently, scientists discovered shocked quartz around the periphery of the dome. Formed in the laboratory only under conditions simulating hypervelocity impact pressures, shocked quartz has a deformed crystalline structure that is strong evidence of meteor impact. Once geoscientists demonstrated it could be an impact structure, Upheaval Dome was further classed as a complex meteor crater due to its central, uplifted core surrounded by a circular valley. In contrast, Meteor Crater in Arizona (see geo-site 4) is classed as a simple meteor crater.

Estimates of the timing of impact range anywhere from 60 to 30 million years ago. In the years since, erosion has reduced the elevation of this structure by some 1,000 feet, removing all telltale evidence of meteor fragments and exposing the deeply fractured rock at the center of the structure. A spur trail from the halfway point of the 8-mile Syncline Loop Trail allows access to the very heart of Upheaval Dome—a locale intimately associated with a colossal collision between Earth and an errant space traveler.

In Canyonlands National Park the presence of a gargantuan circular pit alters the layer-cake simplicity of the surrounding Colorado Plateau.
—Courtesy of Carolyn B. Marks, Richmond, VA

The core of Upheaval Dome is constructed of near-vertical, greenish White Rim sandstone. In contrast, flat-lying rocks of the Colorado Plateau fill the horizon.
—Courtesy of Neal Herbert, National Park Service

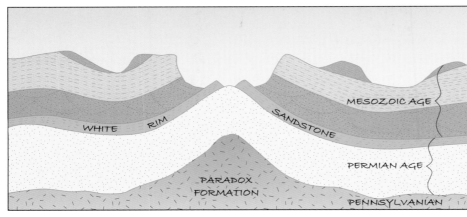

Cross section in support of the meteor impact theory. A central, uplifted core composed of a zone of fractured rock could have formed as flat-lying rock rebounded in response to impact and compression from a meteor.

181

82. Checkerboard Mesa, Utah

37° 13′ 23″ North, 112° 53′ 14″ West
Jurassic Period Wind Deposition

Wind was the deciding factor in the construction of this superlative example of fossilized sand dunes.

Ask any person touring Utah to describe Zion National Park, and the reply would surely include vivid descriptions of smooth, polished, slickrock vistas, towering cliffs, and flat-topped mesas. Ask the same of a knowledgeable geologist, and the reply should simply be Navajo sandstone. This ubiquitous rock unit—the very reason the park exists—makes Zion a visual feast of flushed red, stark white, vibrant orange, and rich copper hues.

About 275 million years ago, the present-day state of Utah was 10 degrees from the equator, inundated by warm seas along the western coast of the supercontinent Pangaea, and drifting northerly at a velocity comparable to the growth rate of the human fingernail. Over time, ocean depths became tidal pools, then floodplains and lakes. Conifer and cycad forests shaded the land. With the dawn of the Jurassic period, another sea inundated the countryside, one of drifting sand. This Navajo Desert effectively covered valley and hill from southern Nevada to central Wyoming. Its geologic descendent is the Navajo sandstone.

The most characteristic aspect of this 2,000-foot-thick sandstone is crossbedding, a feature of sedimentary rocks deposited under the influence of either air or water currents. The presence of several types of terrestrial fossils—land plant, telltale dinosaur footprints, and possibly beetle burrows—indicates wind rather than water deposited the Navajo. The crossbedding in Zion is associated with fossil dunes that constitute the bulk of the rock making up the Navajo sandstone. A dune begins to develop when wind-deposited material accumulates around an obstruction, forming a windward flank. Particles of sand roll up the gentle slope, tumble over the crest, and slump down the leeward slip face. Thus, over a period of time, dunes migrate in the direction of the prevailing air movement as layer upon layer (bed upon bed) of sand is deposited at an angle.

In the Zion area, fluctuating windstorms during Jurassic time altered the orientations of youthful dunes. As a dune was blown in one and then another direction, crossbeds oriented at different angles were deposited, forming a sequence of disoriented crossbeds. Large-scale crossbeds are commonly seen in the Navajo. On an individual basis some are more than 15 feet thick, a measurement suggesting the dunes they were part of may have been 300 feet tall.

Textbook examples of crossbeds are easily accessible at Checkerboard Mesa, along Utah 9, 1 mile west of the eastern entrance to the park. The mesa's name is derived from the checkerboard-like arrangement of vertical fractures intersecting stacked sequences of crossbeds. The fractures formed when the rock expanded after extensive erosion had removed substantial overburden and reduced the pressure on the sandstone. Later, their appearance was enhanced by changes in temperature and moisture conditions.

Fully 90 percent of the sand grains that comprise the Navajo sandstone are quartz, a common rock-forming mineral. During the last stage of transportation from their source area the surfaces of these grains became frosted. This lusterless, ground-glass finish is the result of the particles of quartz rubbing against each other as they were blown around and is yet more proof of the dunes' eolian ancestry. Today, these naked rock models of harmonious form and color continue to bend to the raw power of nature. Time and wind will someday complete their cyclic history, eroding dune rock back to individual sand grains, proving that even—or perhaps especially—in geology, what goes around comes around.

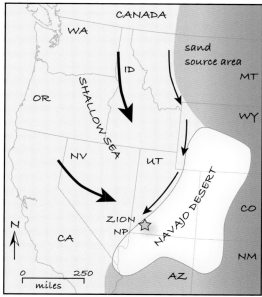

Navajo sandstone sediment was derived from pre-Jurassic-age highlands in Montana and Canada. After longshore currents (small arrows) moved the sand south, offshore winds (large black arrows) drove it inland to the Navajo Desert.

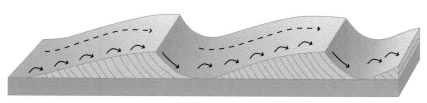

Most dunes are asymmetrical in profile; sand grains bouncing in the prevailing wind direction fuel their migration. Because crossbeds always slope in a downwind direction, they record the flow direction of the air currents that deposited them.

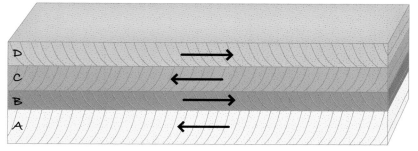

Different orientations and angles of crossbeds (A through D) develop in response to changes in prevailing wind direction (arrows).

Variable arrangements of layered crossbeds, combined with vertically oriented fractures, are the reason for Checkerboard Mesa's moniker.
—Drawing by Patsy Faires, Kernersville, NC

83. San Juan Goosenecks, Utah

37° 10′ 30″ North, 109° 55′ 37″ West
Mesozoic and Cenozoic Era River Erosion

The origin of these tightly wound meanders has long been a classic example of updated evaluation.

The San Juan River, a major Colorado River tributary, forms near Wolf Creek Pass in southwest Colorado. Surrounded by the 14,000-foot-high San Juan Mountains, it begins in true whitewater fashion, changes downstream into cascades, and finally evolves into meanders before entering a drowned valley near Lake Powell in southeast Utah. Because of the tortuous nature of this section of meanders, 5 river miles may equate to less than 2 crow-flight miles. The best vantage point from which to view the heart of this circuitous water route is the overlook at Goosenecks State Park, 4 miles outside the crossroads hamlet of Mexican Hat. Here the surface descends 1,100 feet from the monotony of the surrounding Colorado Plateau to the base of the terraced and twisted canyons that have been hailed as the finest examples of entrenched river meanders in the world. Entrenched meanders are thought to develop when rivers maintain their original meandering courses with little modification as land is uplifted.

Because this desert landscape is nearly denuded of vegetation, the layer-cake nature of the underlying sedimentary rocks is exposed in eye-popping clarity. Two formations compose the canyon walls at Goosenecks State Park: the cliff-forming Paradox of Middle Pennsylvanian age, and the slope-forming Honaker Trail, younger by several million years. In the Paradox Formation isolated lens-shaped fossil coral reefs are interspersed among alternating sequences of limestone, gypsum, and salt. The reefs of this formation have so far produced nearly 500 million barrels of petroleum from the hydrocarbon-rich Paradox Basin, which straddles the Utah-Colorado border northeast of Mexican Hat. In the overlying Honaker Trail Formation, the stratigraphy changes progressively upward from limestone to shale and then sand, suggesting that by the time the shallow sea was depositing this formation it had begun to retreat.

The Goosenecks of the San Juan have a long-standing history of controversy related to the contrarian nature of the silt-laden river. Unlike conventional rivers that conform to local conditions by flowing around topographic obstacles, the San Juan boldly confronts the local geography head-on. Monument Upwarp, a 110-mile-long, 50-mile-wide, convex-upward fold, crosses southeast Utah at a nearly perpendicular angle to the course of the San Juan River. How did this river carve its way across this seemingly insurmountable barrier?

John Wesley Powell, famed for his 1869 exploration of the Colorado River, began the controversy after he floated through Split Mountain Canyon in northeast Utah—a structure similar to the Monument Upwarp—and was mystified as to how "the river . . . cuts into the mountain to its center . . . splitting the mountain ridge for a distance of six miles." He concluded that the Colorado, and by association the San Juan, was older than Split Mountain. Employing their erosive power, these rivers were supposedly able to maintain their already established courses as the land beneath was gradually elevated.

This "old river" theory contrasts with modern thought, which posits the Monument Upwarp anticline formed hundreds of millions of years ago and was subsequently covered by several thousand feet of flat-lying sedimentary rock deposited within river and lake environments. Then, 6 million years or so ago, the San Juan drainage system came into being and began to carve its vertical incision into these younger Cenozoic-age strata, eventually cutting down to older strata and superimposing its course across the trend of the preexisting Monument Upwarp. Science continues to discuss this paradox of time and geology. While the "new river" theory has broad appeal, definitive evidence remains locked within the lazy, looping goosenecks of the San Juan River.

The sinuous Goosenecks are an outstanding example of entrenched meanders.
—Courtesy of Charles Herron, Wilmington, NC

The north-south trend of the Monument Upwarp (dashed line) seemingly forms a topographic barrier to the westward flow of the San Juan River, posing a chicken or egg question: which came first, the river or the anticline?

The canyon of the San Juan is a revealing 3-D display of meander, folded strata, and fossil coral reefs (black). Geoscientists believe that the rate the San Juan has eroded its canyon, as well as that of other rivers of the Colorado Plateau, is similar to the rate the plateau has been uplifted: approximately 1 foot per millennium. (Modified after Stevenson, 2000.)

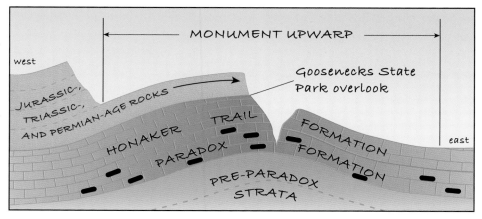

84. Salina Canyon Unconformity, Utah

38° 56' 07" North, 111° 48' 24" West
Phanerozoic Eon Unconformity Development

An American representative of the iconic Scotland geo-site that gave cachet to the science of geology.

Every rock exposure has a story to tell. One speaks of economics, another of the power of mountain building. A third will talk of evidence supporting evolution, while a fourth prattles on about the awesome role of gravity. The litany continues: earthquake catastrophe, climate change cyclicity, meteorite impact, crustal movement, volcanic violence, and sea level fluctuation. The list, in its many permutations, seems endless.

In central Utah, 200 yards north of I-70 at a point 4 miles west of exit 63, east of Salina, is an outcrop ablaze with a panorama of reds and pale reds—the Salina Canyon unconformity. This rock slope tells a tale similar to one recounted along the east coast of Scotland, on a spit of rock immortalized as Siccar Point. At Siccar Point, east of Edinburgh, the science of geology was born, as was the concept of unfathomable time. Here, three men stepped ashore from their rowboat in 1788: James Hutton, a naturalist; John Playfair, a mathematician; and Sir James Hall, a chemist.

Hutton was on a mission to convince his skeptical friends that the rocks at their feet were proof of what was then one of the most heretical claims of science: Earth was far older than the approximate 6,000 years envisioned through analysis of the book of Genesis. In animated fashion, he verbally dissected the jagged shoreline cliff. The base was composed of gray, mica-bearing strata, typical of shale seen throughout the region, except that its layers stood at maximum angle to the horizontal, like books arranged upright on a library shelf. Overlying this sequence was a stack of red sandstone beds, all resting in near-horizontal position. The 90-degree contact of these two units of sedimentary rock was marked by weathered debris derived from the older shale.

Gleefully, Hutton explained the rock disorientation need not be considered the result of short-term catastrophe, as dictated by theology. Instead, the explanation could be as simple as the application of the long-term geologic functions of deposition, compaction, uplift, and erosion, followed by more deposition. Playfair recalled the moment: "The mind seemed to grow giddy by looking so far into the abyss of time." Hutton, today recognized as the father of geology, modestly summarized his presentation: "In the economy of the world I can find no traces of a beginning, no prospect of an end."

The very same historic meeting could have taken place at the Salina Canyon unconformity, where the rocks are identical in configuration and interpretation. In the canyon, horizontal, pastel beds of sandstone and shale of the Paleogene-age Colton Formation, deposited within a floodplain crisscrossed by rivers, overlie vertical, garish, reddish siltstone and sandstone of the Jurassic-age Twist Gulch Formation, deposited in a shallow sea that lapped upon fields of coastal sand dunes. Like Siccar Point, this is a classic example of an angular unconformity—a case where younger sedimentary rock rests upon the eroded surface of folded older rock, providing incontrovertible evidence of a break in the time record.

With the advent of radiometric age dating, it became possible to define the time gaps represented by the rocks at Siccar Point and Salina Canyon as, respectively, 50 and 90 million years. In both cases, missing pages of geologic history echo a logic that eventually allowed scientists to conduct research free from biblical consideration.

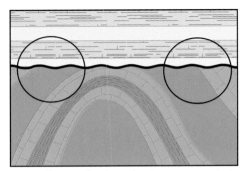

4. Sediment is deposited on the eroded surface, creating an angular unconformity (black line). The photographs (right) show contacts similar to those marked by the circles.

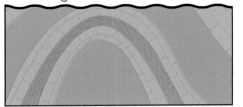

3. Folded rocks are beveled by erosion.

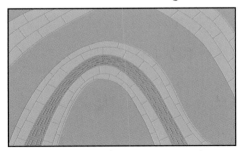

2. Compacted sedimentary rock is folded by mountain building forces.

1. Sediment is deposited and gradually compacted as it is buried.

Angular unconformities are created through a four-stage process of deposition, uplift, erosion, and more deposition. Their common occurrence is evidence of numerous worldwide cycles of tectonics and erosion.

The angular interplay of the two suites of rock at Siccar Point serves as proof that Earth is, in the words of James Hutton, "unfathomably old."
—Courtesy of Dr. Clifford E. Ford, The University of Edinburgh, Scotland

The Salina Canyon panorama of vertical and near-horizontal strata mimics the unconformity defining the scenery of Siccar Point.

85. Bingham Stock, Utah

40° 31' 24" North, 112° 09' 03" West
Paleogene Period Mineralization

This richest hole on Earth is a paradigm for efficient mineral exploration on a worldwide basis.

Rocks, like many books, shed important light on Earth's history. Both, though, must be opened before their cache of information can be revealed. One requires only the turn of a page; the other often depends on excavation as small as a chip from a geologist's hammer or as large as a commercial mine. Nowhere is this better illustrated than at the Bingham Canyon Mine, the industrial operation associated with the Bingham Stock. Described as the world's largest human excavation and one of the planet's most productive mines, this open pit measures 2.75 miles wide by 4,000 feet deep. Located 25 miles southwest of downtown Salt Lake City, it is often described as the only manmade feature visible to the naked eye from a space shuttle. The visitors center is open seven days a week, April through October.

To date, the Bingham mine has produced more than 15 million tons of copper—more than any other mine in history. In addition, these Oquirrh Mountain rocks have yielded a bonanza of valuable by-product metals: some 620 tons of gold, 5,000 tons of silver, 625 million pounds of molybdenum, and significant amounts of lead and zinc. This production dates to 1863 with the discovery of copper-bearing veins exposed along the slopes of Bingham Canyon. The geologic story, however, goes back much further.

The foundation of north-central Utah is composed of sandstone and limestone deposited in a sea that invaded the region during the Pennsylvanian period. These sedimentary strata remained undisturbed until a multitude of mountain building forces began to form the Oquirrh Mountains. Starting some 40 million years ago, the intrusion of massive volumes of molten rock, accompanied by equal volumes of chemically active hydrothermal fluids (hot, watery solutions that escape from cooling magma), accentuated the closing phase of this intense cycle of folding and faulting. A significant portion of this intrusive activity was centered on the Bingham Canyon area, the birthplace of the Bingham Stock, a world-famous pluton.

As both the magma and hydrothermal solutions began to cool, a range of ore minerals crystallized from the magma and precipitated from the solutions. Some minerals were left throughout the granite rock of the Bingham Stock, principally in the form of molybdenum- and copper-bearing material, which was scattered as tiny grains or filled numerous fractures. This form of mineralization is termed disseminated.

Then the hydrothermal fluids interacted chemically with sedimentary rocks the magma had intruded, forming additional ore deposits that enveloped the stock in an approximately 1-mile-wide zone. The innermost zone is composed of minerals containing copper, sulfur, and iron, whereas the outermost is composed of rocks rich in lead, zinc, silver, and smaller amounts of gold. Once the Bingham Stock had been emplaced, erosion gradually began to expose its upper portions to the atmosphere, creating an environment in which oxidation enriched surface ores. The mine removed these ores first, which were three times richer in minerals than those now being extracted.

The Bingham Canyon Mine has long been considered a classic and model deposit for metal exploration programs worldwide. The level of copper concentration being mined today can be as high as 0.63 percent, the equivalent of 13 pounds of malleable metal per ton of ore. An estimated 900 million tons of copper-bearing ore remain encased in the rocks of this yawning open pit. With average consumption of 30 pounds per person per year, these reserves should keep America self-sufficient when it comes to copper for decades.

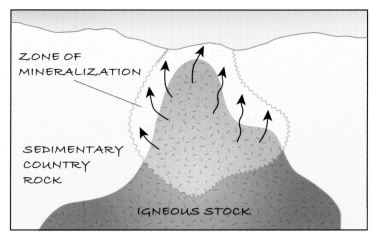

Hydrothermal fluids initially precipitated minerals within fractures of the stock and then within the adjacent country rock. Arrows represent the movement of the fluids.

The tallest building in the United States, the 110-story Willis Tower (formerly known as Sears Tower), would reach but one-third of the way to the rim of the Bingham Canyon pit. —Photograph © by John Meneely, Phoenix, AZ

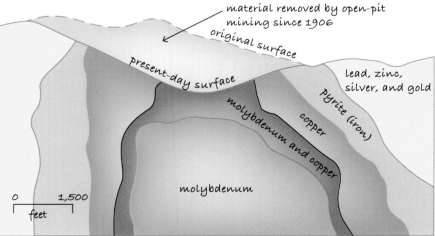

Cross section of the ore zones (labeled) in both the stock (outlined in black) and the adjacent sedimentary rock.

189

86. Whipstock Hill, Vermont

42° 53' 53" North, 73° 16' 01" West
Ordovician Period Mountain Building

A spectacular roadside outcrop offers insight into the adolescence of the Appalachian Mountains.

Vermont 279, informally known as the Bennington Bypass, officially opened in 2004, conveying traffic westward to New York State. Construction unearthed a sensational exposure of Early Ordovician–age Bascom Formation marble at Whipstock Hill, 0.5 mile inside Vermont on the right-hand side (east) of the road. Composed of an anticline bookended by two synclines, this outcrop presents a dramatic glimpse into the gargantuan geologic powers responsible for the early evolution of the Appalachian Mountains.

The 500-million-year-old marble of Whipstock Hill, and the rest of Vermont's terrain, has been a long time in the making. Between 1,300 and 1,000 million years ago the landmass Grenvillia collided with Laurentia, creating the Grenville Mountains. For several hundred million years the now-enlarged Laurentia (proto–North American continent) experienced first the ravishing effects of erosion and then the even more landscape-altering forces of rifting—similar to the deep-seated, pull-apart pressures at work today in the East African Rift System. When invasive marine waters inundated the sagging topography, the Iapetus Ocean, predecessor to the Atlantic, was born.

As the sun rose on the first years of the Paleozoic era, Vermont lay submerged within the warm, equatorial environs of the Iapetus. This get-go moment of Appalachian history emerged as rivers draining the highlands to the west were transporting copious amounts of sand to these shallow waters. As the sea gradually invaded the land, carbonates began to blanket the landscape. This sandstone-limestone suite constitutes the bedrock foundation of the Vermont Valley, a steep-walled physiographic province that separates the Green Mountains from the Taconic Mountains and is home to Whipstock Hill. Similar rocks make up the entire stretch of the Appalachians.

By Middle Ordovician time, the dimensions of the Iapetus Ocean, having advanced to a mature state of development, began to decline in accordance with the plate tectonics theory: oceans open and then they close. Compression replaced extension as the dominant tectonic force, only to be followed by subduction and the inevitable formation of an island arc. This arc, similar in geometry and constitution to the present-day Aleutian Islands archipelago in the Pacific, moved westward on a tectonic conveyer belt in the upper layers of Earth's mantle. As the arc converged on the eastern shore of Laurentia—at a velocity similar to the growth rate of the human fingernail—the western extremities of the Iapetus Ocean were squeezed shut. Vermont became completely high and dry, and has remained so since.

This event folded and metamorphosed layers of rock, including the carbonates deposited in the Iapetus, which became the Bascom Formation marble, then thrust them westward along the Maple Hill Fault—up to a distance of 60 miles—in a quest for topographic glory. That goal was achieved in the creation of both the Taconic Mountains and the Green Mountains, the latter the rock-ribbed, central backbone of Vermont. This phase of compressional tectonics created multitudes of anticlines and synclines, including the folded features exposed at Whipstock Hill.

In one final cataclysm of mountain building, another emigrant landmass—Avalonia—crashed into the east coast of Laurentia 60 million years later, uplifting Vermont yet again and giving the marble of Whipstock Hill one final jolt of compression. The land now known as Vermont was complete, fully assembled and in possession of a rugged, mountainous profile. In retrospect, it seems there has never been a moment within the last 500 million years when this state has lacked a mountainous profile. However, just as nature abhors a vacuum, it dislikes elevation, and over the ensuing years erosion and weathering have reduced once lofty highlands to the rolling hills of today. The contorted rocks of Whipstock Hill bear grand witness to this lengthy history of orogeny.

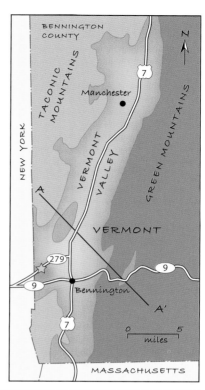

The Whipstock Hill roadcut (star) lies adjacent to Vermont 279.

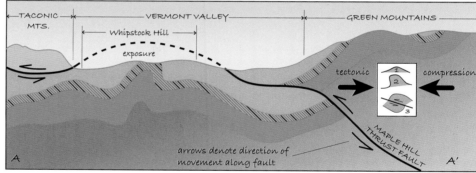

Cross section through Bennington County (see figure above for location) showing how this region was folded and faulted during the early part of the Paleozoic era. The inset diagram demonstrates the progression from anticline to thrust fault caused by tectonic compression.

Near-vertical battleship gray layers of Bascom Formation marble (center) mark the transition from an anticline (left) to a syncline at the Whipstock Hill roadcut.

87. Great Falls, Virginia

38° 59' 45" North, 77° 15' 12" West
Quaternary Period Waterfall Retreat

This cataract of foaming, churning water has been in retreat for the last 35,000 years.

From the discovery of the New World until mid-seventeenth century, transoceanic explorers made it a point to investigate every principal river entering the Atlantic Ocean, often looking for the fabled Northwest Passage. More often than not, a waterfall that signaled the upstream limit to river navigation thwarted their search. Turning obstruction to advantage, American colonists built villages by these cascades and used whitewater power as a source of energy. In time, these settlements developed into metropolitan areas that today share a common geologic heritage: they are fall-line cities.

The fall line—a contact between upland and lowland—of the east coast marks the boundary between two physiographic provinces: the Piedmont Province and the Atlantic Coastal Plain. The Piedmont Province is traceable from New Jersey to central Alabama, extends inland to the Blue Ridge Mountains, and reaches a maximum width of 300 miles in North Carolina. This plateau is composed of a collage of igneous and metamorphic terranes constructed over an interval of millions of years as the supercontinent Pangaea was being assembled. That amalgamation ended approximately 250 million years ago.

Later, the shoreline created by the breakup of Pangaea marked for a time the eastern border of the Piedmont. During the Cretaceous and Cenozoic ages, however, river systems draining the Appalachian Mountains, west of and adjacent to the Piedmont, deposited sediments that extended the shoreline eastward and created the flat-lying Atlantic Coastal Plain. Because rivers erode these younger sedimentary strata much more readily than they do the very resistant crystalline bedrock of the Piedmont, a series of waterfalls and rapids often identifies the contact of these provinces—the fall line. One of the most scenic of these waterfalls, Great Falls, is located at Great Falls Park along the Potomac River, 15 miles upstream from Washington DC.

The geologic history of Great Falls and its relationship to the fall line dates from that time when river systems began to drain the slopes of the youthful Appalachian Mountains. With the arrival of Pleistocene glaciation, sea levels dropped, causing the ancestral Potomac River to extend its course and its flow to be reinvigorated. As a result, its valley deepened, exposing the juxtaposition of the Piedmont Province and Atlantic Coastal Plain. The fall line was born. Today, a section of this contact bisects the 88-acre Theodore Roosevelt Island, lying across the Potomac River from the Watergate complex in Washington DC. Here, the fall line waterfalls of the Potomac River began life some 35,000 years ago. Since then, however, those early rapids have slowly retreated upstream to Great Falls, so they lie a full 15 miles from their starting point. This recession rate has in recent years averaged 2 feet per year, evidence of the tremendous power of turbulent, flowing water and its ability to erode the very hardest of rock.

Now protected by the National Park Service, Great Falls is undoubtedly the most dramatic display of agitated and tumbling water found along the Potomac River. The nearly 1,000-foot-wide channel abruptly narrows here to an average of 80 feet and drops 76 feet in elevation as the river flushes along a series of rapids and cascades before entering Mather Gorge, a mile-long canyon formed by near-vertical walls of late Proterozoic–age metamorphic rock. If knowledge of the past is any entry point into the future, this upstream migration from its starting point at the fall line, fueled by the power of water in motion, will continue until that distant day when Great Falls reaches the lofty slopes of the Blue Ridge Mountains and transitions from spectacular rapids to electrifying cascades.

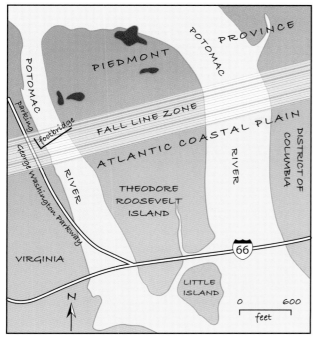

The geologic contact between the Piedmont Province and the Atlantic Coastal Plain crosses Theodore Roosevelt Island in Washington DC. Exposures of Precambrian-age schist (black) on the north portion of the island identify the Piedmont Province. The absence of Precambrian rock on the south half identify the Atlantic Coastal Plain.

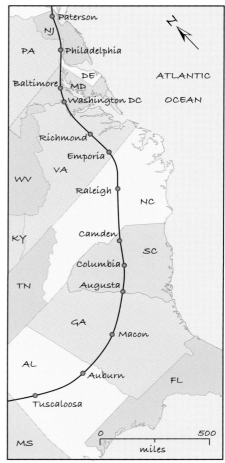

A series of riverfront cities define the trace of the fall line (heavy black line) along the east coast of the United States. Each city was founded at the contact of the Piedmont Province (left of line) with the Atlantic Coastal Plain (right of line). This contact generally marks the effective limit of upstream navigation. Washington DC is an exception.

The foaming, sensational rapids at Great Falls Park are considered the steepest and most sensational of any seen along the fall line rivers draining the eastern United States.

88. Natural Bridge, Virginia

37° 37′ 38″ North, 79° 32′ 33″ West
Cenozoic Era River Erosion

This historic wonder of nature was caused by river convergence, capture, and eventual subsurface flow.

The Commonwealth of Virginia is composed of a central valley (part of the much larger Great Appalachian Valley) snuggled between the Blue Ridge Mountains and the Valley and Ridge Province. Because the folded and fractured limestone and dolomite strata that underlie large sections of the Great Appalachian Valley are susceptible to dissolution, the valley's distinguishing characteristic is karst, topography marked by numerous sinkholes and caves. Perforated terrain of this nature makes up as much as 20 percent of Earth's land surface and a resounding 25 percent of the United States. Karst extends over portions of twenty-nine of Virginia's ninety-five counties, creating more than 4,300 caves and close to 49,000 sinkholes.

During the early years of the Paleozoic era, the eastern portion of proto–North America was basking in an extended period of quiescence. The dominant geologic event centered on a continuing rise in sea level, causing a widespread invasion of the continental interior. Geologists have named this salt-water inundation the Sauk Sea; it was a subportion of the much larger Iapetus Ocean—precursor to the Atlantic Ocean. These warm oceanic waters deposited thick sequences of carbonate sediment, which later indurated into limestone and dolomite rock formations. Several of these, known by the names Copper Ridge, Elbrook, Chepultepec, and Beekmantown, make up the principle bedrock of the Great Appalachian Valley and its related karst topography.

Perhaps the most intriguing of the many karst landmarks of Virginia is Natural Bridge, 2 miles east of exit 175 off I-81 in Rockbridge County. The vital statistics of this historic and most famous natural bridge in America, if not in the world, are impressive. The top of the bridge is 215 feet above the landscape below. Its 90-foot-long span is 50 to 100 feet wide and 50 feet thick and contains more than 450,000 cubic feet of rock. Some ingenious individual has even calculated that its massive volume of light-gray Beekmantown dolomite and bluish gray Chepultepec limestone—both determined to be some 470 million years old—would weigh in at 36,000 tons, give or take a few pounds.

In 1794 Thomas Jefferson, the bridge's first owner, described it as a "convulsion of nature." Today, however, there is general agreement that both surface and subterranean aspects of river erosion closely associated with the history of Cedar Creek, the stream that placidly flows under the arch, created Natural Bridge. Several million years ago Cedar Creek, while still in the act of finalizing its drainage pattern, merged with neighboring Cascade Creek by way of a great meander loop. Later, however, an elongated sinkhole that had developed nearby in the carbonate bedrock captured Cedar Creek and diverted it into a subterranean course. With the passage of time dissolution extended this captive drainage, the final effect being the creation of a tunnel as well as a redirected Cedar Creek. Most of the tunnel roof long ago collapsed, leaving Natural Bridge as evidence of this remarkable sequence of events.

In the not too distant geologic future this segment will also crumble to rubble, most probably the result of the continuing freeze-thaw process, in which water freezes in cracks and wedges rock apart. In the final tally, erosion and weathering are the ultimate enemies of all exposed rock.

An 1839 sketch of Natural Bridge differs little from the same view today, evidence that the forces of erosion are quite slow as gauged by the life span of the average individual. —Courtesy of Historic Images, Auburn, CA

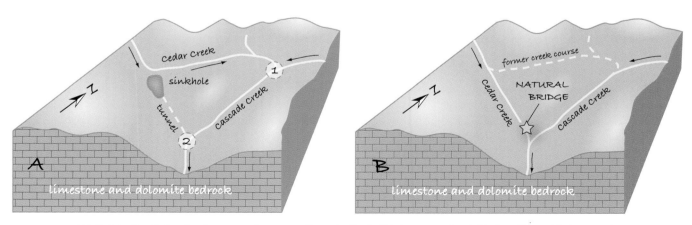

(A) Originally, Cedar Creek was a tributary to Cascade Creek (1), but a nearby sinkhole redirected it into a subterranean drainage where it joined Cascade Creek farther downstream (2). (B) Ultimately, the tunnel forming the redirected river course collapsed, leaving a small remnant as Natural Bridge. Arrows show stream flow direction.

89. Millbrig Ashfall, Virginia

36° 41′ 39″ North, 83° 17′ 05″ West
Ordovician Period Volcanic Eruption

A supervolcano unleashed its pent-up fury in perhaps the largest eastern-U.S. eruption in the last 500 million years.

Earth is unique among planets. Seventy-two percent of its surface is enveloped in water, and the entire planet is wrapped within an all-encompassing atmosphere. Each of its mega-environments—land, sea, and air—swarms with life that simultaneously undergoes extinction and evolution. At an age of 4,600 million years, it remains geologically healthy as evidenced by its heartbeat and pulse rate—both earthquake activity and volcanic eruptions.

Approximately 1,500 active volcanoes mottle Earth's surface—the acne of its terrestrial complexion. Of the roughly 500 that have erupted within recorded history, more than half define the Ring of Fire, the chain of volcanism extending around the periphery of the Pacific Ocean. Many of these displayed Vesuvian characteristics: violent expulsions of viscous, gas-filled lava accompanied by large volumes of fine-grained ash. Since 1982, seismologists have used the Volcanic Explosivity Index, a 0-to-8 scale wherein each number represents a tenfold increase in the volume of extruded pyroclastic material, to measure the magnitude of Vesuvian-type eruptions. Seven such eruptions are of particular significance.

Mount St. Helens erupted for a full nine hours. Pinatubo, in the Philippines, caused average global temperatures to decline a full 1 degree Fahrenheit. The eruption of Vesuvius in AD 79 killed an estimated 10,000 to 25,000 people and buried the cities of Pompeii and Herculaneum. Heard fully 3,000 miles away, the explosion of Krakatoa in Indonesia activated barographs (used for measuring barometric pressure) around the world. The ash Tambora (Indonesia) blew into the atmosphere created a worldwide famine. Long Valley reconfigured the California landscape, and the collapse of the Yellowstone Caldera deposited ash throughout the western United States.

These acknowledged behemoths are impressive in scope and destructive power, but their reputations are affected when compared to the violent eruptive event that occurred when the volcanic island arc Taconica collided with proto–North America. Of the more than sixty known volcanic ash beds that resulted from this Late Ordovician event, one possesses impressive eruptive credentials. The 454-million-year-old Millbrig "Big Bentonite" bed was deposited across an estimated 965,000 square miles in a layer that ranges from 3 to 6.5 feet thick. In the United States it can be traced from Vermont to Nebraska and Mississippi and northward into Ontario.

The Millbrig ash bed increases in thickness to the southeast. Because a larger amount of ash would have settled to the ground closer to the volcano, this configuration suggests the parent volcano was sited in the general vicinity of present-day Kentucky, in an area then inundated by the proto–Atlantic Ocean. The explosion of this unnamed supervolcano belched more than 270 cubic miles of ash into the atmosphere, giving it a Volcanic Explosivity Index rating of 8, the upper limit of the scale.

A beautiful 3-to-6-foot-thick exposure of Millbrig ash is part of a hillside abutting railroad tracks in a remote part of Lee County. Travel 10 miles west of Jonesville on US 58 and turn north on Virginia 946/621 (Burning Well Road). Drive 0.1 mile and turn west on Virginia 621 and proceed 0.6 mile. Drive or walk up the rough dirt road on the right for 2 blocks to a small parking area. Relatively small, the exposure is 60 yards to the right along the railroad tracks. This reddish band of semiconsolidated rock, bookended by gray-toned units of stratified limestone, attained superstar status when it was declared evidence of possibly the largest known eruption to have occurred in the eastern United States during the last 500 million years. A new chapter in geologic, as well as human, history would be necessary should an eruption of this intensity ever occur again.

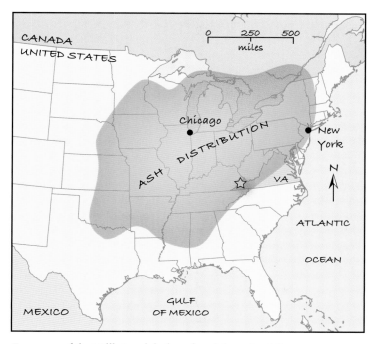

Exposures of the Millbrig ash bed are found throughout the eastern and midwestern United States. This geo-site (star) is located near the southeastern edge of the ash distribution. (Modified after Kolata et al, 1996.)

Ordovician-age limestone developed by the warm waters of the proto–Atlantic Ocean envelop the 3-to-6-foot-thick Millbrig ash bed (red exposure) in Lee County.

VOLCANO	DATE	DEATHS *ESTIMATED	VOLUME	VOLCANIC EXPLOSIVITY INDEX
Mount St. Helens, USA	1980	57	0.67	4-5
Pinatubo, Philippines	1991	300-900 *	1.2-2.5	5-6
Vesuvius, Italy	AD 79	10,000-25,000 *	2.5-25	6
Krakatoa, Indonesia	1883	36,417	6	6
Tambora, Indonesia	1815	>70,000 *	35-50	6-7
Long Valley, USA	0.76 ma	—	150	7
Yellowstone, USA	2.1 ma	—	585	8

Statistical review of the eruptions of seven famous volcanoes (ma = millions of years ago). Volume is in terms of cubic miles of erupted pyroclastic material.

90. Catoctin Greenstone, Virginia

38° 01' 57" North, 78° 51' 32" West
Proterozoic Eon–Paleozoic Era Continental Evolution

A cycle of supercontinental birth, growth, and death encapsulated in a sequence of metamorphosed lava beds.

The stories that some rocks contain are like the stories that some books tell. They can be succinct and uncluttered, or they can wander in and out of descriptive passages designed to confuse the reader. The Catoctin greenstone story is a straightforward tale of only two events, each having a major impact on the development of the North American continent. This geologic vignette can best be read by studying a fenced-off roadcut about 20 miles west of Charlottesville. It is visible from the sidewalk of the I-64 bridge at exit 99, which connects Skyline Drive with the Blue Ridge Parkway. To the seasoned traveler, this geo-site is known as Rockfish Gap.

EVENT 1

A supercontinent is born and dissected, then oceans form.

A billion years ago the assemblage of the North America craton—the geologic heart of the continent—was well under way. With the advent of the Grenville orogeny, the ancestral heartland of Africa became tectonically welded to the craton, significantly increasing its size and forming mountains. Shortly thereafter, the addition of a handful of migrant mini-continents gave the map of Earth the appearance of one large slab of land, called Rodinia, surrounded by a universal body of salt water. At that time the future state of Virginia lay very close to the center of this newborn supercontinent.

For several hundred million years, a period of calm prevailed, highlighted by the abrasive forces of weathering and erosion. The Grenville Mountains were worn away. Rodinia remained featureless and devoid of life, and a hot spot was pressing hard against its underbelly, causing it to stretch and then fracture. This process of rifting created numerous fissures through which an estimated 1,000 cubic miles of magma poured onto the surface during periodic eruptions that took place some 700 million years ago. Between eruptions rivers deposited sand on the cooled lava. The resulting Catoctin Formation, composed of multiple flows of basalt lava interspersed with layers of sandstone, blanketed a corridor that today extends from North Carolina to Newfoundland.

As rifting slowly divided Rodinia, the zone of separation, or rift, collapsed and began to fill with salt water. Earth then became a two-ocean planet. Further disassembly transposed Virginia from inland to seaside real estate.

EVENT 2

An ocean is destroyed and a new supercontinent is born.

In geology as in life, what goes around comes around. In the continuing episode of the supercontinent cycle, during which multiple masses of land assemble and separate on average every 500 million years, both oceans were strangled as the scattered pieces of Rodinia—minicontinents—began their own process of amalgamation. When proto-Africa collided with proto–North America, the Appalachians formed and became the final tectonic act in the construction of a new supercontinent, called Pangaea.

The Catoctin basalt flows, situated dead center in this collision, were altered to greenstone, a metamorphic rock composed of pistachio green epidote and forest green chlorite. Simultaneously, the interspersed sedimentary rocks were compressed into bands of boudinaged sandstone—sausage-shaped, elongate structures created by excessive physical deformation. The Catoctin greenstone, truly a tour de force rock, encapsulates the birth, growth, destruction, and assemblage of supercontinents in a colorful manner. It's an interesting read at this revealing geo-site, especially for those having time only for the abbreviated version.

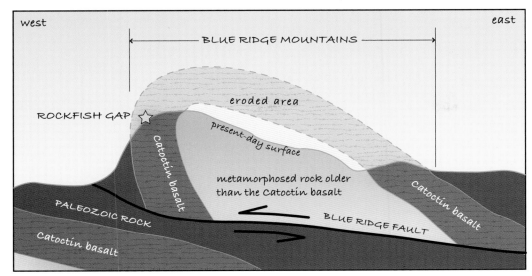

During the Appalachian orogeny the Catoctin basalt was folded and metamorphosed into greenstone. The eroded west limb of this gargantuan anticline is exposed at Rockfish Gap. Arrows denote direction of movement along the fault. (Modified after Bentley, 2008.)

The process of metamorphism altered the original black hues of the Catoctin basalt to shades of green, as seen in this example, thus the name greenstone. Pencil for scale.

Along with the metamorphism of the Catoctin lava flows, interbedded layers of sandstone were stretched and thinned into sausage-shaped boudins (outlined).

91. Mount St. Helens, Washington

46° 12' 00" North, 122° 11' 21" West
Holocene Epoch Volcanic Eruption

This maelstrom of volcanic destruction is a precursor to future catastrophic Cascade Range events.

Before her day of infamy, Mount St. Helens was regarded as the smallest and prettiest of the big, snowcapped stratovolcanoes that make up the igneous spine of the state of Washington. Perhaps because her recorded history of activity exceeds that of any other volcano in the conterminous United States, this is the most analyzed of the chain of monster eruptive landforms that constitute the Cascade Range of the Pacific Northwest.

At precisely 8:32:17 a.m. local time, May 18, 1980, subterranean forces triggered by a magnitude 5.1 earthquake burst through the flank of Mount St. Helens, spewing forth a 680-degree-Fahrenheit stew of gas, water, and exploding rock. The resultant release of energy was the equivalent of 21,000 exploding Hiroshima-type atomic bombs. A trio of gargantuan landslides quickly coalesced into a singular avalanche, reportedly the largest in recorded history, that moved an estimated 0.7 cubic mile of rock debris down the valley of the North Fork of the Toutle River at speeds of 80 miles per hour. Simultaneously, two columns of depressurized gas shot skyward, converged at 80,000 feet, turned day into night, and deposited an estimated 0.67 cubic mile of gray ash across southwestern Washington and ten states to the east. When the eruption ended nine hours later, 23 square miles of material had been removed from the mountain and 1,300 feet of the summit of this now violent poster child of the volcanic world were missing, dropping her from fifth to eighty-seventh place on the list of the state's highest mountains. Trees were felled throughout a 230-square-mile-area, fifty-seven people lost their lives, and a devastating chapter had been written in the volcanic annals of Earth.

Whenever a major eruption occurs, the question rises anew: how many active volcanoes are there? The consensus is about 1,500, with *active* referring to those eruptions occurring within the past 10,000 years. Among these, fifty to seventy boil over every year and at any point in time—now is as good as any—some twenty are in the process of actively proving their eruptive prowess. These statistics are for terrestrial eruptions only; estimates of the number of volcanoes lurking within the depths of the world's oceans exceed one million.

Since the eruption of 1980, the scientific community has put much effort into developing a comprehensive, long-range eruption predictability program, with some success. While obstacles remain, because volcanoes behave differently and each has its own lifestyle, the quest continues, driven by the volcano-related deaths of some 850 people worldwide each year.

At Johnston Ridge Observatory, at the end of Washington 504, 52 miles east of exit 49 (Castle Rock) on I-5, geology in action is the focus. The viewing plaza, sited in the heart of the blast zone, overlooks a stunning 180-degree panorama of the ruptured crater, the landslide deposit, and a still-steaming, 1,000-foot-high lava dome. Inside, interpretative displays and a 3-D volcano model illuminated by optical fiber light technology tell the story of May 18. Outside, strategically placed monitoring equipment continuously records seismic activity, changes in heat generation, and ground deformation—respectively the heartbeat, temperature, and weight fluctuation of the volcano—initiating a new era in the science of eruption forecasting. The future of Mount St. Helens is rife with scientific speculation. If her past is any indication, however, she stands today a relatively docile damsel waiting with certitude for her next day of infamy.

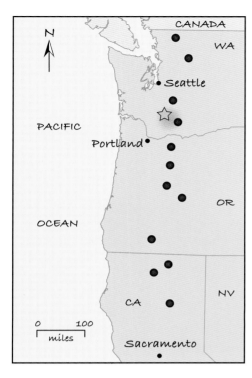

On average, two of the thirteen volcanoes of the Cascade Range erupt every century, Mount St. Helens (star) most recently.

Before the eruption.
—Courtesy of the U.S. Geological Survey Cascades Volcano Observatory

After the eruption, a smoldering lava dome rose from the reincarnated throat of Mount St. Helens.
—Courtesy of the U.S. Geological Survey Cascades Volcano Observatory

92. Dry Falls, Washington

47° 36′ 25″ North, 119° 21′ 49″ West
Quaternary Period Flood Erosion

The greatest-ever flood crafted a scabland topography of channel, canyon, plunge pool, and cataract.

The northwest sector of the contiguous United States features scenery that is at once geographically beautiful and geologically controversial. Descriptions of the area demand a generous use of superlatives: the largest floods, the largest lake ever created by an ice dam, and one of the greatest waterfalls in geologic history. What sounds like hyperbole is, in fact, definitive. After decades of debate and denial, the story of the Channeled Scablands, the 3,000-square-mile region centered on eastern Washington, preserves unconventional but irrefutable evidence of the awesome power of flowing water. The overlook at Sun Lake–Dry Falls State Park, adjacent to Washington 17, 7 miles southwest of Coulee City, offers insight into the scablands and one of its distinctive features: Dry Falls.

First to study the scablands, geologist J. Harlen Bretz traversed thousands of miles of foreboding terrain in the 1920s before concluding that catastrophic flooding had sculpted the basalt bedrock into a pockmarked collection of buttes, mesas, and coulees. When he suggested the flooding happened within weeks, if not days, he was accused of the heresy of catastrophism. Not only had he failed to identify the source of the flood, he was suggesting a short-term, cataclysmic cause at a time when the geologic community was entrenched in the logic of uniformitarianism—the belief in the cumulative effect of subtle actions over extended periods of time. When he learned that a massive lake existed in Montana during the most recent ice age, Bretz rearranged his evidence. Again he was labeled, at best, a maverick. Finally, aerial photographic evidence turned the tide and restored his reputation. After years of rejection and debate, the true story of the scablands could be told.

At the height of the Pleistocene epoch, a finger of ice formed a dam across the Clark Fork River, giving birth to Glacial Lake Missoula in Montana. At maturity the lake inundated 3,000 square miles, rose to a level of 4,350 feet, as witnessed by the lakeshore scars on Mount Sentinel in Missoula, and contained 500 cubic miles of water, about the combined volume of Lakes Erie and Ontario. As hydrostatic pressure increased, dam failure became inevitable. Like champagne exploding out of an uncorked bottle, Lake Missoula burst through its icy restraint at a rate of 10 cubic miles per hour, two hundred times the Mississippi River rate at flood stage and as much as ten times the combined flow of all the rivers on Earth.

This thundering slurry of calved icebergs, soil, and 200-ton boulders bulldozed across western Washington at speeds exceeding 60 miles per hour. Anastomosing currents gouged and chiseled humongous potholes (see geo-site 34), plunge pools, and cataracts. The most awe-inspiring feature created by the flood is Dry Falls, a 4-mile-wide, 400-foot-high bulwark—three times the height of Niagara Falls—constructed of multiple Miocene-age basalt flows and capped by coulees more than 10 feet deep.

Researchers eventually discovered the "Missoula flood" was but one of more than eighty that ravaged the scablands over a period of 3,000 years. At the age of ninety-six, Bretz accepted the Penrose Medal, the highest award given by the Geological Society of America, for contributions to geological knowledge and his commanding role in shaking the roots of dogmatic uniformitarianism. He later reportedly told his son, "All my enemies are dead, so I have no one to gloat over."

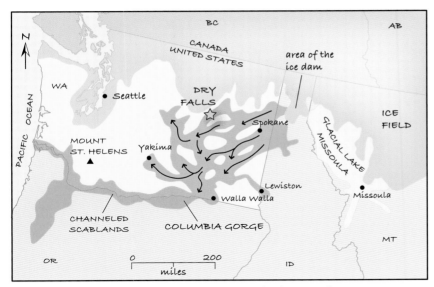

The key elements of the Dry Falls story are Glacial Lake Missoula, the ice dam of northern Idaho, and the scablands of Washington and Oregon. Arrows denote the direction floodwaters flowed. (Modified after Alt and Hyndman, 1984.)

This massive chunk of weathered bedrock is one of many transported 100 miles or more by the surging floodwaters of Glacial Lake Missoula.

The roof of Dry Falls is incised by a cluster of coulees, the alignment of which parallel the direction the sculpting waters flowed. A 0.5-mile-wide, deepwater plunge pool—Dry Falls Lake—inundates the base of the former waterfall.

93. Seneca Rocks, West Virginia

38° 50′ 02″ North, 79° 22′ 54″ West
Permian Period Mountain Building

Tall sentinels of fin-and-crag topography identify the wall of a decapitated anticline.

The geology of the Appalachian Mountains has conceivably been studied to greater extent than that of any other mountainous region of the world. In 1803 in France, Count Constantin-François Volney published the first American geologic map, covering the area from Canada to Florida and from the Atlantic Ocean to the Mississippi River, but it was minimally distributed. Six years later, William McClure presented the first readily available map of the geology east of the Mississippi, using colors to distinguish classes of rock. By 1830 interest in American geology was widespread, especially as it related to the raw materials needed by an ever-growing nation. Much was expected from the hills and valleys bordering the eastern coast of the United States.

Today the Appalachian Mountains are viewed as a region of classic field studies from which many theories have sprung, ranging from plate tectonics to the origin and distribution of crude oil. Generations of geologists have combed these ancient hills because of the great thicknesses of sedimentary and volcanic rock of Paleozoic age. If all the different formations of this age in the Appalachians were stacked one on another, the resulting pile would measure more than 35,000 feet. Contrast this with the less than 3,000 feet of similar rock found overlying the geologic basement throughout the Mississippi River watershed.

The Appalachians comprise four northeast-southwest-trending physiographic provinces. The easternmost province is the Piedmont, a plateau composed of crystalline, igneous and metamorphic rock that millions of years of erosion has reduced to nearly peneplain status. Next is the Blue Ridge, a high-relief alignment of Precambrian- and Paleozoic-age rocks, some as old as 1 billion years; folding and faulting have displaced bedrock of this province as much as 10 miles to the northwest from its original location. Then comes the Valley and Ridge, a roller-coaster landscape of anticlines and synclines dominated by thick sequences of sedimentary rock. And the westernmost of the provinces is the Appalachian Plateau, a region of gently folded columns of Paleozoic sedimentary rock that decrease in thickness to the west and contain vast resources of coal, crude oil, and natural gas.

Deep in the mountainous terrain of Pendleton County, two of these physiographic provinces abut each other along a highland ridge known as the Allegheny Front. Southeast of this front, Valley and Ridge rocks are severely folded and faulted. To the northwest, strata of the same age are affected only by the style of minor folding characteristic of the Appalachian Plateau. The viewing platform at Seneca Rocks is an optimal place to envision the topographic characteristics that differentiate these two provinces. Seneca Rocks, off West Virginia 28, is roughly 22 miles southwest of Petersburg. Here, mountain building forces, which affected the region as recently as 260 million years ago, compressed the 300-foot-thick, Silurian-age Tuscarora quartzite into a tight, convex-upward fold several miles in width. In the process this dense and resistant formation went from its original horizontal attitude to one of more than 90 degrees. The Wills Mountain anticline, as it is known, defines the most northwesterly boundary of the large-scale anticline-syncline-folded terrain that constitutes the Valley and Ridge Province. The geology immediately inland from this anticline changes abruptly to the gently folded landscape of the Appalachian Plateau. Forest growth hides this change from the view on top of Seneca Rocks.

Over time, as erosion breached the crown of the Wills Mountain anticline and thereby exposed much of the underlying Ordovician-age limestone, Seneca Rocks, composed of tough, erosion-resistant Tuscarora quartzite, emerged. Positioned on the northwest flank of the eroded anticline, the rocks were reluctant to bow to the forces of weathering and erosion, resulting in the development of a classic example of fin-and-crag topography—narrow, steeply tilted plates of rock. Gunsight Notch, a deeply eroded vertical fracture, accentuates the center of this outcrop. This fin, crag, and gap "gemstone" is the most spectacular rock exposure in West Virginia.

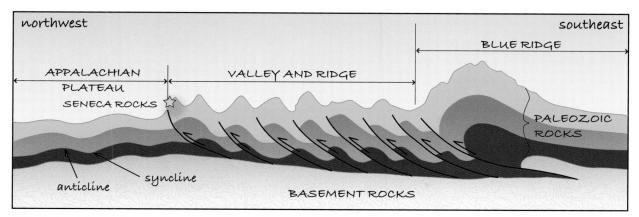

Disturbed and distorted rocks of the Valley and Ridge Province join gently folded strata of the Appalachian Plateau along the Allegheny Front. This tectonic union is clearly defined at Seneca Rocks. Arrows denote direction of movement along faults.

Reputed as the finest example of fin-and-crag topography in the central Appalachian Mountains, Seneca Rocks literally stand on end, enveloping Gunsight Notch.

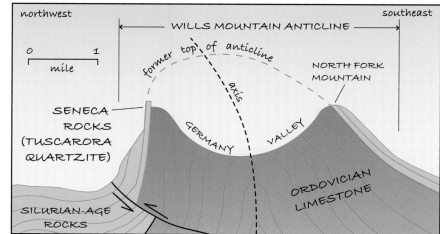

Cross section of the Wills Mountain anticline. The bend of the axis is evidence that the compressive forces that formed this massive fold were directed from the southeast.

94. Roche-A-Cri Mound, Wisconsin

44° 00′ 10″ North, 89° 49′ 08″ West
Phanerozoic Eon River and Wind Erosion

A mesa and butte terrain spared from the all-consuming destructive effects of continental glaciation.

During the Pleistocene epoch, mile-thick tongues of ice possessing the capacity to reduce anything standing in the way to sand and dust repeatedly overwhelmed the northern portion of the contiguous United States. The present-day courses of the Missouri and Ohio rivers mark the southernmost reach of these invasions. For reasons still subject to debate, a 14,500-square-mile area centered on southwestern Wisconsin escaped these repeated onslaughts of frozen water. Various glacial tongues bypassed this Denmark-sized region in a gigantic pincerlike military style of advance, then rejoined forces to the south and enveloped it within a wall of ice. Today this is called the Driftless Area because it lacks the ground-cover debris (glacial drift) recognized as the footprint of glacial movement. Numerous topographic sentinels, such as chimneys, crags, towers, pinnacles, mesas, and buttes, provide further proof that this area has never been subjected to the run-amok process of glaciation.

A group of these outliers is gathered within the "castellated mound" region, a descriptive Wisconsin-speak phrase meaning any castlelike, isolated hill offering a good view of the countryside. The photographic pearl of the collection is Roche-A-Cri, a 300-foot-high sandstone prominence crowned by an overlook that offers a 120-mile perspective in all directions. It forms the geologic heart of Roche-A-Cri State Park, about 1.5 miles north of the community of Friendship, off Wisconsin 13 in Adams County.

The tale begins some 500 million years ago when rivers meandering sluggishly through braided channels distributed a sizable thickness of quartz-rich sand across the flat terrain of Wisconsin. Later, when a shallow, inland sea invaded the region, the depositional style changed to limestone and its magnesium-rich cousin, dolomite. This era of calcium carbonate deposition continued into the Silurian and Devonian periods and then ended approximately 375 million years ago.

Thereafter, the geological history of Wisconsin is unfortunately incomplete. There is no evidence that sediments were deposited during the multimillion-year segment of time lasting from the Devonian period until the initiation of the Pleistocene ice age. Throughout this segment of time, however, river and wind erosion deeply incised the existing, ancient sedimentary bedrock, leaving the countryside peppered with thousands of landforms similar in both configuration and composition to Roche-A-Cri. Then some 2 million years ago the glaciers arrived, grinding away most of these pre–ice age landforms and heavily modifying the rest. The exceptions were those mounds sheltered within the haven of the Driftless Area.

Roche-A-Cri, described as the steepest and the most conspicuous hill in Wisconsin, is representative of all the remaining castellated mounds of the state. Lying 12 miles from the eastern edge of the Driftless Area, this elongate, Cambrian-age sandstone rampart was for a time an island in the center of Lake Wisconsin, a meltwater glacial lake that during its prime approximated the size of the present-day Great Salt Lake.

Today, Lake Wisconsin is gone, but Roche-A-Cri lives on, resisting the relentless forces of wind and water. However, like all Wisconsin mounds, it is ephemeral, and in the near geological future, perhaps several million years from now, it too will be subdued. Such are the whims and ways of nature, or stated more succinctly: deposition giveth, and erosion taketh away.

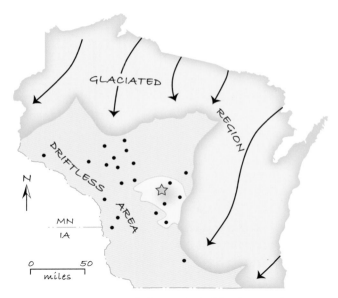

Wisconsin is composed of two principal landscapes: the glaciated region and the Driftless Area, the latter also known as the castellated mound region. Roche-A-Cri (star) is located near the center of the former Lake Wisconsin (light blue). Arrows represent direction of ice flow, and black dots other castellated mounds.

The view from the top of Roche-A-Cri is punctuated by at least six sister castellated mounds that help define the Driftless Area.

Ship Rock, northeast of Roche-A-Cri along Wisconsin 13, is a castellated mound in an advanced state of physical deterioration.

95. Van Hise Rock, Wisconsin

43° 29' 21" North, 89° 54' 56" West
Precambrian Time Folding and Metamorphism

A solitary column of quartzite and slate reveals the secrets of an ancient era of mountain building.

In the early days of the twentieth century, the interpretation of Precambrian rocks was difficult—and for good reason. By definition more than 542 million years old, these crones had in all probability been subjected to all forms of geologic onslaught, not the least of which are the tortures of metamorphism, during which excessive heat and pressure alter the original texture and structure of a rock mass. Metamorphism is to rock like fire and flood are to the pages of a book: both make the reading thereof troublesome.

The confusion over interpreting this era began to change when the geologic community recognized the importance of a 14-foot-high, 6-foot-wide outcrop that stands along the east side of Wisconsin 136, 1 mile north of Rock Springs in Sauk County. In 1997 the National Park Service named this pillar of Precambrian-age metamorphosed sandstone and shale Van Hise Rock, in honor of the geologist who used it to identify the changes that occur in rocks subjected to the forces of mountain building.

Charles R. Van Hise, first recipient of a PhD in geology from the University of Wisconsin and later president of the same institution, also used this monolith to illustrate the relationship of metamorphism to large, folded sequences of rock in Wisconsin. While he conducted much of his pioneering fieldwork throughout the Precambrian terrain of the iron and copper mining districts surrounding Lake Superior, he frequently probed the ancient rocks of the Baraboo District because they were closer to his home in Madison.

The topography of the Baraboo District centers on a pair of ridges—North Range and South Range—that are mainly shaped by exposures of Baraboo quartzite, an erosion-resistant rock that began its career as beach sand 1,700 million years ago. Later, younger sediments buried these same sands and associated layers of mud, which were then compacted, crumpled into a series of folds, and finally metamorphosed. The Baraboo rocks were first discussed in 1858 in a state of Wisconsin report, in which Edward Daniels described them simply as "lofty ranges of hard quartzite." Two decades later new research suggested they might be exposures of the same rock body, the layers of which dip to the north as low as 15 degrees on the South Range and as high as 90 degrees on the North Range. Then, in 1904, one of Van Hise's students, using field criteria developed by the master himself, correctly identified the regional structure as a syncline, a concave-upward fold of rock.

Standing center stage on the North Range, Van Hise Rock boasts a central, dark, slanted display that is a classic example of inclined slaty cleavage. Slaty cleavage develops in fine-grained rocks enveloped by coarser-grained rocks as tectonic activity bends them into folds. Minerals in the fine-grained rock move microscopically during this process, aligning in one direction. The result is that the rock, when exposed at the surface, can break, or cleave, along the planes of aligned minerals. The slabs of highly fractured, coarser-grained, pink quartzite surrounding the slaty cleavage of Van Hise Rock contain shadowy evidence of crossbedding, a feature either air or water currents formed as they rearranged the original layering of sand grains that later were metamorphosed to quartzite.

Properly interpreted, the elements on display in the Van Hise Rock—layers of quartzite surrounding slaty cleavage—give true structural definition to the rocks throughout the Baraboo District. The angular relationship between slaty cleavage and the vertical strata of quartzite is a clue to the geometry of the regional syncline underlying the Baraboo Valley. The syncline formed some 1,630 million years ago by tectonic plate collision, a date determined by radiometric age dating. Van Hise used this Precambrian-age exposure to demonstrate how important one small exposure can be to the overall understanding of a region, a case of the microworld mimicking the macroworld. No wonder this site continues to be a mecca for students of geology from around the world.

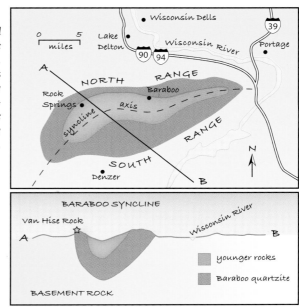

Geologic map and cross section of the Baraboo syncline. Geologists realized its true nature when they understood the relevance of small-scale features, such as slaty cleavage.

The physical characteristics of Van Hise Rock indicate this isolated outcrop has evolved from sand and shale to mountain fold. —Drawing by Patsy Faires, Kernersville, NC

Van Hise Rock.

96. Amnicon Falls, Wisconsin

46° 36' 39" North, 91° 53' 29" West
Proterozoic Eon Continental Rifting

A fossil earthquake zone marks one megafracture that failed to subdivide a youthful North America.

More than 1 billion years ago proto–North America was a developing continent on the geologic go, energized by plate collisions that were welding microcontinents together to form the North American craton, a vast terrane now identified by igneous and metamorphic rock of Precambrian age. The girth of this craton was nearly three-quarters the size of present-day North America.

While this landmass amalgamation was under way, a hot spot—a plume of molten mantle rock—had formed deep within Earth's interior. This thermal nozzle was aimed upward to a site that would one day be the center of Lake Superior. Because continental crust is an excellent insulator, the rising plume of rock met resistance and spread laterally along the base of the craton's crust. Like taffy being pulled, the craton slowly thinned and then cracked. Once formed, ruptures quickly extended to the southwest and southeast and did not die out until they reached what is today the basement rock underlying Kansas and Ohio. The 1,000-mile-long Midcontinent Rift had been born. The East Africa Rift System is a modern example of such a rift.

From day one, the Midcontinent Rift was literally on a tear, intent on splitting ancestral North America into a series of subcontinents. Under continuing attack by pull-apart forces, the ruptures grew ever wider, and infusions of molten rock, their curtain-of-fire eruptions causing night-sky incandescence, vomited upward and quickly filled each newly created void. For 22 million years this process continued unabated—crack and fill, crack and fill—until the rift had widened to 120 miles. Eventually, its molten roots crystallized into a 10-mile-thick column of basalt that sealed the ruptures, and the rift's axis was distinguished by a very large and well-defined graben, an elongate trough bounded by faults. Here and there, freshwater lakes filled deep depressions, adding a sense of relief to the otherwise barren scenery. By this time, however, the Midcontinent Rift had exhausted its fifteen geologic minutes of fame. A northwesterly roving, 250-mile-wide, 2,000-mile-long microcontinent called Grenvillia was beginning to make fender-bender contact with the craton.

Extension forces flipped to those of compression, squeezing the central graben upward into a horst, an elongate, uplifted block of rock bounded by faults. The rift had received a fatal tectonic blow, but the structural integrity of the craton was saved. Had extension continued, Duluth, Minnesota, and its sister city Superior, Wisconsin, might possibly be separated today by a salt-water gulf equal to the expanse of the Atlantic Ocean between Lisbon, Portugal, and New York City.

Paleozoic sedimentary rock and glacial drift almost entirely cover the ancient remnants of the Midcontinent Rift, but one quite illustrative exposure can be seen upstream from the covered bridge in Amnicon Falls State Park, about 14 miles south of Superior and off US 2. Here, Amnicon Falls tumbles over erosion-resistant Precambrian-age lava beds, which are juxtaposed against younger, less-resistant Precambrian-age sandstone at the base of the falls. The contact between the basalt and sandstone is a fault plane that identifies the northwest edge of the horst that formed in the core of the Midcontinent Rift. One billion years ago, earthquakes caused the basalt to rise along the fault relative to the sandstone. Overall, it rose 8,000 feet and is dramatic proof of the tremendous amount of compressive force generated when Grenvillia collided with ancestral North America.

The Midcontinent Rift remains today possibly the world's best example of an aborted attempt at continental bifurcation and definitive proof that the forces of geology, once unleashed, do not always continue to a successful conclusion.

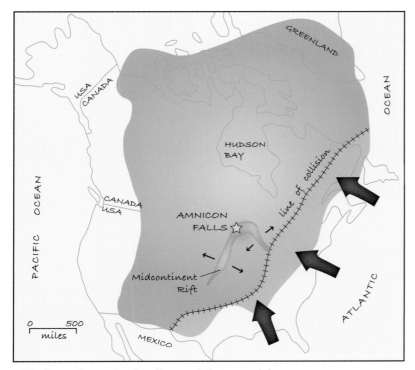

The forces of extension (small arrows) that created the Midcontinent Rift were overwhelmed when the errant microcontinent Grenvillia (blue) collided head-on (large arrows) with the North American craton (green).

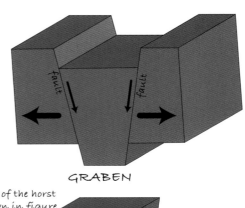

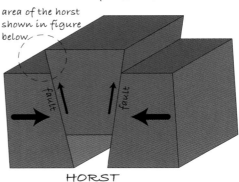

The graben nature of the Midcontinent Rift was reversed to that of horst when tectonic extension was replaced by compression (large arrows). The small arrows denote direction of relative movement along faults.

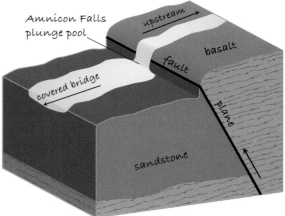

Amnicon Falls plunges over the faulted edge of the horst that is the structural core of the Midcontinent Rift. Here, billion-year-old basalt abuts younger sandstone. —Courtesy of Bonnie Gruber, Wisconsin Department of Natural Resources

211

97. Green River, Wyoming

41° 32' 40" North, 109° 28' 59" West
Eocene Epoch Petroleum Generation

A looming mile marker of the Overland Trail may contain the answer to the energy problems that face America.

In an era when America's dependence on foreign sources of petroleum is recognized as critical, the prescience of M. King Hubbert, the first geologist to investigate the geopolitics of oil field depletion, is being lauded under the rubric "He told us so." His prediction that domestic oil production would peak in about 1970 was accurate to within a matter of months. As a prognosticator, Dr. Hubbert hit the bull's-eye—or did he?

Since the Age of Petroleum began in 1859, the uneven geographic distribution of this resource has caused political and economic turmoil worldwide. The Middle East today contains some two-thirds of the world's proven reserves of petroleum. Saudi Arabia alone has control of 20 percent of world reserves. In contrast, at the beginning of the twenty-first century, the United States could claim only 2 percent. To a nation wed to an oil economy, these statistics are daunting. And yet, if they are broadened to include unconventional petroleum reserves—those not present in rock in the liquid state—the American outlook would improve exponentially.

Much of the dusty desert terrain of eastern Utah, western Colorado, and southwestern Wyoming is composed of fine-grained sedimentary rock of the Green River shale, which was deposited within a series of lakes that blanketed the region 50 million years ago. Year in and year out, immense volumes of barely altered remains of aquatic life and blue-green algae sank into the oxygen-starved and murky depths of these tepid alkaline waters, forming extensive layers of organic muck separated by bands of silt. Once buried and exposed to elevated temperature and pressure, the organic content of the clay-sized sediment changed into kerogen, a solid, waxy, bituminous precursor to petroleum.

Sited off exit 89 of I-80, 0.5 mile west of the community of Green River, Tollgate Rock is a classic and colorful example of this earth-tone rock. Here, as in nearby exposures, the darker layers have the richest kerogen content. A sliver of rock broken from the highest-quality layer, the jet black Mahogany Ridge zone, will sustain a flame when ignited.

Investigation of the Green River shale has been under way for more than a century. The conventional method of production involves multiple steps: surface mining, crushing, heating to 700 degrees Fahrenheit, vaporization, condensation, and finally refining. Economics and degradation of the environment are continuing concerns, but research continues because the potential resource gain is staggering. It is estimated the Green River shale contains the equivalent of as much as 1.8 trillion barrels of unconventional oil—more than all the petroleum produced since the dawn of the Age of Petroleum. Of this volume, 800 billion barrels of oil—three times the liquid petroleum reserves of Saudi Arabia—is deemed recoverable. The U.S. Geological Survey estimates this formation alone contains a full 50 percent of the world's supply of in-place shale oil.

The desert sands of the Arabian Peninsula may presently hold the first five numbers to the international petroleum reserve lottery, but if and when concerns of cost and environment are overcome, the Green River shale beds of the western United States may harbor the all-important Powerball number that wins the jackpot.

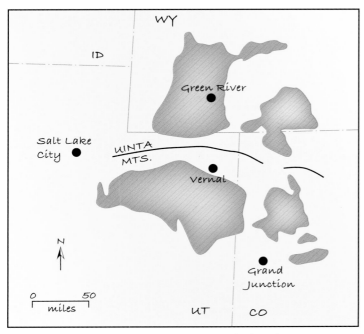

Thick kerogen-bearing beds of Green River Shale (green) were deposited within a series of western lakes during the Eocene epoch. The shale was deposited over 17,000 square miles.

When a fresh face of the rock of the Mahogany Ridge zone is exposed, its black gold richness becomes apparent. This zone could yield 60 gallons of oil per ton of rock. Six-inch ruler for scale.
—Courtesy of Charles Herron, Wilmington, NC

Earth-tone strata crowning Tollgate Rock are richer in kerogen than the lighter-toned rock of its base.

98. Devils Tower, Wyoming

44° 35' 24" North, 104° 42' 53" West
Paleocene Epoch Magma Influx

Debate continues as to the exact nature of this Hollywood icon: is it intrusive or extrusive in origin?

If the sculptured images of Mount Rushmore could turn their heads 180 degrees and peer 80 miles to the west, eight granite, presidential eyeballs would see a mammoth, 5,112-foot-high crystalline sentinel intersecting the horizon. Hailed by explorers, fur traders, and settlers as the most recognizable feature of the northern Great Plains, revered by many Native Americans as *Mateo Tepee*, the "Grizzly Bear Lodge," and recognized today as Devils Tower, this steep-sided, igneous prominence surges 867 feet above the grassland and ponderosa pine landscape of northeastern Wyoming.

Perhaps its most striking feature is the flat-topped upper framework, constructed of a series of spectacular vertical rock columns that descend onto a blotched, circular bench that oozes onto the countryside. Most of these polygonal columns have five surfaces; others are four- to seven-sided. The largest measure more than 10 feet in diameter at their base, tapering to half that at the top. Bundled together, they are perhaps the finest, most massive, and tallest example of columnar structure on Earth.

Columnar structure is a phenomenon associated with lava flows and magma intrusions (see geo-site 9). The columns develop as molten rock cools and crystallizes from its surface inward, creating an environment of tension that cracks the solidifying rock into clusters of generally vertical fractures. Under ideal circumstances, the fractures intersect to form six-sided columns. In nature, however, fluctuations in heat and chemistry often create columns with three to more than six sides.

The base of this sky-penetrating edifice is composed of a field of boulders up to 10 feet in diameter. This apron of talus is caused by rockfall, a form of mass wasting that occurs frequently in mountainous terrains through the repeated freezing and thawing of water that accumulates in the cracks of rock. Each time water freezes in a crack, it wedges the rock on either side farther apart until blocks of rock break away from the main mass.

Sixty million years ago, mountain building associated with the formation of the modern Rocky Mountains uplifted western North America. As part of this activity, magma intruded the basement rock of northeastern Wyoming. Controversy abounds over what happened next in the case of Devils Tower. One argument suggests the magma reached the surface and formed a volcano—an extrusive event—and Devils Tower is a plug, a pipelike mass of igneous rock that represents the main conduit of a former volcanic vent. But if this is truly a weathered plug, associated ash beds or lava flows should be evident in the vicinity. None are.

The second theory posits that magma sputtered out at depth—an intrusive event—and then moved laterally, forming a laccolith, a flat-floored mass that intrudes between two layers of sedimentary rock and arches overlying strata. After it crystallized, erosion stripped away the overlying sedimentary layers, leaving a massive core of igneous material standing high and mighty. The bulbous base is certainly reminiscent of the overall shape of a conventional laccolith, and in matters of texture, mineralogy, and style of jointing, the rock is more typical of an intrusive origin than one of extrusion. Not surprisingly, this theory is more widely accepted.

Structure and size, form and presence, history and controversy. No wonder Theodore Roosevelt, one of the Mount Rushmore presidents, selected this geo-site as the first national monument in 1906, and seven decades later Steven Spielberg chose it as the landing pad for extraterrestrials in his award-winning movie *Close Encounters of the Third Kind*.

The 220 rock-climbing routes of Devils Tower are close in count to the many columns of igneous rock that characterize its surface. —Courtesy of the National Park Service

Rockfall is the primary manner in which talus slopes are built. The field of debris that surrounds the bulbous base of Devils Tower is typical of talus slopes found at the base of steep rock outcrops throughout the Rocky Mountains.

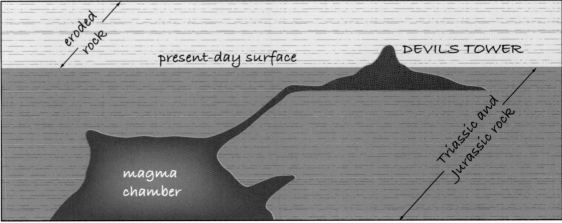

Cross section in support of an intrusive origin. Devils Tower was possibly born when the erosion of overlying sedimentary rock exposed a laccolith.

99. Fossil Butte, Wyoming

41° 50' 12" North, 110° 46' 16" West
Paleogene Period Fossilization

In a High Plains setting, a panorama of diverse subtropical life is reincarnated in rich detail.

Forty-eight million years ago the first rays of the morning sun crossed the eastern shore of Fossil Lake in the section of North America that would someday become southwestern Wyoming. Dragonflies flitted from lily pad to cattail and mosquitoes swarmed. Luxuriant growths of cypress, fig, and palm trees, interspersed with patches of fern, shaded the shoreline. Farther afield, groves of oak, willow, and maple harbored various birds and mammals, among them plant-eating tapirs and four-toed horses.

Along the horizon, fir and spruce enveloped the mountain slopes, covering rocky spurs in which colonies of bats sought daytime shelter. Within the lake, 15-foot-long, predatory crocodiles lazily cruised the depths, sharing with shallow-water-loving crocodiles the apex of a diverse food chain composed of snakes, turtles, insects, stingrays, snails, shrimp, and at least twenty species of fish ranging in size up to the 6-foot-long gar. In short, normalcy reigned, but not for long. The wet, subtropical, frostless, Amazon Basin–like environment of Fossil Lake was on the cusp of a climate inversion that would transform it into an elevated, sagebrush-covered steppe characterized by low rainfall and extended, bitter winters.

Today the sedimentary rock remnants of this earlier time period are recognized as one of the most outstanding depositories in the world of preserved freshwater fauna and flora of Paleogene age. These Green River Formation strata are so rich in fossils that paleontologists commonly refer to them as a *lagerstätten*, a collection of fossils distinguished for its diversity or quality of presentation—in this case both. The heart of this treasure trove lies within the buff-colored strata forming the upper 300 feet of Fossil Butte, the signature topographic feature of Fossil Butte National Monument, 15 miles west of the town of Kemmerer.

The environment of Fossil Lake was ideal for it to become a fossil factory. Sluggish streams deposited fine-grained sediment along the shoreline and adjacent shallows. Offshore, the chemical-rich water yielded a constant "snowfall" of calcium carbonate that settled to the lake floor as ooze. This blanket of physical and chemical sediment was the perfect tomb for the timeless incarceration of any fauna and flora that settled on its surface.

The precision of fossilization here almost defies description. A bat, its wings folded in a position of death, is preserved to the extent that undigested food remains visible in its digestive tract. The structure of insect antennae and the delicate veins in the wings of a dragonfly cannot be differentiated from living versions. Fish feature perfectly preserved skin, scales, and teeth.

While most of these plants and animals died of natural causes, frequent occurrences of mass mortality events are recorded in layers that contain literally hundreds of fossilized fish within a 10-square-foot slab of rock—not in disarticulated distribution, but as if an entire school of fish suddenly sank to the lake floor. Research has yet to identify the cause, but hypotheses center on either an occasional bloom of blue-green algae poisoning of the water or sudden changes in temperature or salinity. Today, Fossil Butte offers world-class geologic evidence that, while one cannot go home again in the sense of Thomas Wolfe, one can here go back in time to that prolific Paleogene period, when humans had yet to gain dominance in the lineup of life on Earth.

Some twenty-five, up to 8-inch-long skeletons of Knightia, *an extinct member of the herring family, are contained in this sample of Green River Formation rock.* —Courtesy of the National Park Service

Dragonflies, mosquito larvae, water striders, snails, and crayfish are ample evidence of the freshwater nature of Fossil Lake. This dragonfly's wingspan is 2.4 inches. —Courtesy of the National Park Service

Fossil Butte is unequaled in the world as a depository of a diverse assemblage of Paleogene-age flora and fauna.

100. Steamboat Geyser, Wyoming

44° 43' 25" North, 110° 42' 12" West
Neogene and Quaternary Period Volcanic Activity

Superimposed on the heart of a supervolcano, this cauldron of steam is fueled by the pulsations of a hot spot.

In an 1,800-square-mile region centered on the northwestern corner of Wyoming, the Yellowstone supervolcano encompasses more than half of the world's geysers, hot springs, and fumaroles, including Steamboat Geyser—currently Earth's active geyser with the tallest eruption of boiling water and steam. It is also the site of Yellowstone National Park, the most volcanically active region in the contiguous forty-eight states. Over the past 2 million years, multiple titanic eruptions have occurred here. The first ejected an estimated 585 cubic miles of rubble and ash, a volume almost a thousand times greater than that attributed to the eruption of Mount St. Helens. This ratio of thousand to one is why the Yellowstone area is classed a supervolcano.

While its history has been traced back 17 million years, the supervolcano's origin is wrapped within a shroud of opposing theories, two of which dominate geologic discussion. Did day one dawn because of the cataclysmic impact of a meteorite, or when fractures that extended to the outer depths of the mantle violently rifted Earth's crust? Either way, a localized sector of the mantle instantly liquefied and morphed into the Yellowstone hot spot, a subsurface plume of molten mantle rock over which a field of volcanic activity developed.

When the hot spot was born, the westward drift of the North American Plate was well under way, encouraged by the continuing breakup of the supercontinent Pangaea. As continental rock slowly moved over and across the static, subsurface position of the Yellowstone hot spot, an expansive streak of igneous activity—marked by volcanoes, ashfall, and lava flows—tattooed the surface. The result is the Snake River Plain of southern Idaho (see geo-site 22).

The latest of the three "recent" Yellowstone eruptions occurred 640,000 years ago, creating an estimated 240-cubic-mile void within the parent magma chamber. Because nature abhors a vacuum, the volcanic cone readily collapsed into the void, leaving a topographic depression measuring 29 by 45 miles, the approximate size of the state of Rhode Island, centered in Yellowstone National Park. Called the Yellowstone Caldera, this depression exists over the evolving, double-pronged hot spot, as youthful and scalding as ever and poised a mere 3 miles below the surface, like an upright, double-barreled shotgun ready to be triggered once again.

Steamboat Geyser is only one of the numerous manifestations of thermal activity that cluster around and within the Yellowstone Caldera. Its credentials are impressive, but its reputation is iffy. It is the centerpiece of perhaps the hottest and most changeable thermal basin—Norris Geyser Basin—in the world. Here, water with a record high temperature of 459 degrees Fahrenheit was recorded in a well. A typical eruption of Steamboat Geyser can shear limbs from adjacent lodgepole pines, blast columns of boiling water more than 300 vertical feet, and follow up with a steam display rising to 500 feet and lasting a day or more. On the other hand, it is notoriously erratic in its periods of activity: intervals between major eruptions range from four days to fifty years.

Steamboat Geyser is strikingly symbolic of the Yellowstone supervolcano, a collage of simmering, spitting, and whistling volcanic teapots within and around a caldera that functions like a giant pressure cooker. These thermal features shout loud and clear that this hot spot geo-site is still as much alive today as it was yesterday.

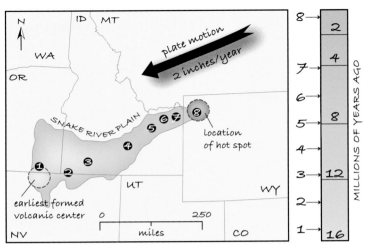

The westward-migrating North American Plate is tattooed with a string of volcanic centers (1 through 8, old to young) forming the Snake River Plain, all the result of the Yellowstone hot spot, which now lies below northwestern Wyoming.

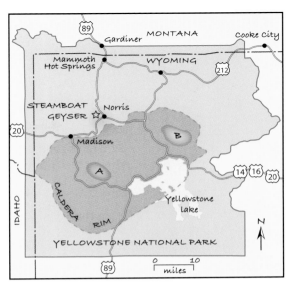

Resurgent domes are areas of topographic swelling due to hot rock and magma beneath the surface. The domes (A and B) within the Yellowstone Caldera rose some 20 inches during the mid-twentieth century, evidence this hot spot is still very much alive.

The only geyser eruptions that exceeded the 300-foot-high Steamboat record were those of Waimangu Geyser in New Zealand between 1900 and 1904. In November of 1904 this activity stopped as suddenly as it had started, leaving Steamboat (seen here) the world's tallest geyser.

101. Specimen Ridge, Wyoming

44° 52' 12" North, 110° 17' 47" West
Eocene Epoch Fossilization

Entire forests were destroyed and entombed one upon the other during an extended period of volcanism.

For any Yellowstone National Park official, the date of August 20 should conjure up dark memories of that Saturday in 1988 when fires that had been devouring the park since late June doubled in size. By the time the final flames were extinguished two months later, 1,875 blackened square miles of the greater Yellowstone region identified an event of unprecedented magnitude.

Even though these events were reported in exhaustive detail, one point of interest remained unnoticed. As devastating as the big burn was, it decimated only one of the more than two dozen forests that extend across the region. These unscathed forests remain steadfastly immune to incineration because the trees are petrified. Once living and viable, these trees, stacked in forest upon forest, anchor steep slopes in the greater Yellowstone region in a state both stony and sterile. One such slope is Specimen Ridge, accessible 4 miles east of Tower Junction in Yellowstone National Park.

Some 55 to 40 million years ago, a giant reservoir of molten rock (unrelated to the region's current volcanic activity) lay beneath the Yellowstone area. Eruptions that were both intermittent and explosive spewed forth from an estimated thirteen interconnected volcanic vents. During that period of volcanism the countryside was covered with a 5,000-foot-thick blanket of ash and fragmented volcanic rock. Today, these vent areas are hidden within the fields of lava and sedimentary deposits of igneous debris that define the wilds of the Absaroka Mountains, a horseshoe-shaped range of light-colored rock east and southeast of Yellowstone Lake.

The ancient ash eruptions of those volcanoes fell on terrains shaded by growths of walnut, maple, oak, redwood, pine, chestnut, magnolia, dogwood, and sycamore, plus lesser stands of birch, hickory, and elm, all nourished by a temperate climate and watered annually by some 50 inches of precipitation. The eruptions buried the lower portions of the trunks, killing the trees and leaving the upper sections to decay. The stumps are found today rooted in tuff-rich sandstone, interpreted to be immature soil horizons. Thick, overlying layers of conglomerate were deposited in lahars, flows of mud, rock, and woody debris that raced down the sides of the volcanoes.

Migrating groundwater dissolved silica from the acidic soils in the region and later deposited it in the pores of the now-dead trees (see geo-site 40). When the woody structure decayed, the growth rings and microscopic cellular features of many trees remained as impregnated mineral matter. Over and over, this process repeated itself until an estimated twenty-seven fossil forests—count them, twenty-seven—lay stacked one upon the other, a 1,400-foot-thick record of cyclical growth, death, and preservation. In one well-studied sector of Specimen Ridge, 70 percent of the fossil trees are found in an erect (upright) position; the remaining percentage are prone.

Periodically, someone questions whether the forests at this geo-site are in situ, that is, in their original place of growth. Analyses of fossil remains from elsewhere in Yellowstone National Park present evidence—disturbed soil profiles, lack of fossilized bark and limbs, and the absence of correlation between wood and pollen types—of an allochthonous state, which means the forests were moved from their original position. For example, the weight of a massive lahar could have dislodged a section of forest floor and moved it downslope. Studies of the fossil forests on Specimen Ridge, however, show no evidence that they are allochthonous. The presence of undisturbed root systems and lahar deposits that flowed around and buried the trunks strongly supports the designation of this site, deemed one of the largest stone forests in the world, as in situ. Specimen Ridge fossils are also outstanding examples of the tenacity of the biologic world, as seen by the ability of the forests to continually reestablish themselves in spite of an extended and destructive reign of volcanism. Such persistence in the face of change is unusual within the world of fossil plants.

The organic details of numerous tree rings in this Specimen Ridge fossil were preserved by the infiltration of silica-enriched groundwater. —Courtesy of the National Park Service

This fossilized redwood, anatomically indistinguishable from redwoods growing today in California, stands in situ 1.5 miles east of Tower Junction.

Petrified stumps complete with fossilized root systems remain anchored in a sedimentary rock composed of volcanic debris high on the slopes of Specimen Ridge. —Courtesy of the National Park Service

Glossary

Acadian orogeny. The second of the three Paleozoic-age phases of mountain building that were instrumental in the development of the Appalachian Mountains.

Alleghanian orogeny. The third of the three Paleozoic-age phases of mountain building that were instrumental in the development of the Appalachian Mountains.

angular unconformity. A contact where younger rock rests upon the irregular surface of older rock with bedding that is at a different angle because it was folded prior to the deposition of the younger rock. The contact represents a gap in the geologic record—a period of erosion or a time during which no sediment was accumulating.

anhydrite. A mineral composed of anhydrous (no water) calcium sulfate.

anticline. A convex-upward fold in layered rock commonly caused by compressive tectonic forces.

Appalachian Basin. A linear trough in the eastern United States that is 1,100 miles long in a northeast-southwest direction and on average 160 miles wide in a northwest-southeast direction. It contains Paleozoic-age sequences of sedimentary rock, thousands of feet thick, that were ultimately folded and faulted into the Appalachian Mountains.

arête. A sharp-edged ridge sculpted by glaciers.

ash. Extremely fine-grained material derived from a volcanic eruption

asthenosphere. The weak rock layer of the Earth in which rock can flow like taffy; the part of the upper mantle that underlies the lithosphere.

Avalonia. An island-arc landmass of late Proterozoic to early Paleozoic age that became part of modern-day New England during the building of a supercontinent.

badland. Topography with little or no vegetation and characterized by rugged, high-relief terrain formed within unconsolidated or poorly cemented clay or silt.

Baltica. A late Proterozoic–age microcontinent that collided with Avalonia, and later Laurentia, to form the supercontinent Pangaea. Today it forms the core of eastern and northern Europe.

basement. The undifferentiated complex of rocks, often a combination of igneous and metamorphic types, that underlies the rocks of interest in an area.

Basin and Range Province. A region that extends through much of the southwestern United States and is characterized by numerous north-south-oriented, elongate, arid valleys bounded by mountain ranges.

batholith. A large, intrusive igneous mass, commonly composed of granite, extending 40 square miles or more.

bed. A layer in sedimentary rock that is distinct from the ones above it and below it due to changes in sediment size, color, or composition or other sedimentary characteristics.

bedding plane. The surface of separation between any two distinct beds of sedimentary rock.

boulder. A rock mass larger than a cobble, having a diameter greater than 10 inches.

breccia. A sedimentary or volcanic rock composed of angular fragments encased in a solidified matrix of finer-grained sediment.

bryozoan. A small invertebrate characterized by colonial growth and a calcareous skeleton.

butte. An isolated, flat-topped, steep-sloped hill with a summit smaller in extent than that of a mesa.

calcareous. Consisting of calcium carbonate.

calcite. A common mineral composed of calcium carbonate.

calcium carbonate. A whitish crystalline compound found in limestone, chalk, and many shells.

caldera. A circular depression caused by either the collapse or explosion of a volcano.

caprock. A hard rock layer that is more resistant to erosion and weathering than underlying units.

carbonate. A sedimentary rock composed of appreciable amounts of limestone, chalk, or dolomite.

chalcedony. A semitransparent variety of quartz recognized by its waxlike luster.

chalk. A soft, fine-grained rock composed almost entirely of the calcareous parts of microorganisms or algae.

chert. A hard, fine-grained rock composed entirely of silica.

cinder cone. A conical volcanic feature composed of fragments of igneous rock emitted during an explosive eruption.

cirque. A half-bowl hollow eroded by a mountain glacier.

clay. A particle of sediment less than 0.00016 inch (0.004 millimeter) in diameter.

claystone. A fine-grained sedimentary rock made up of clay minerals and lacking the characteristic thin layering of shale.

coal. A combustible rock formed by the compaction of altered plant remains.

cobble. A rock fragment with a diameter between 2.5 and 10 inches.

conglomerate. A coarse-grained sedimentary rock composed of rounded pebbles, cobbles, and boulders held together by a fine-grained matrix.

continental drift. A general theory postulating the movement of large plates of continental crust across a substratum of oceanic crust. With time and new information this idea evolved into the plate tectonics theory.

core. The central zone of Earth's interior below the depth of 1,800 miles. It is divided into an inner, solid subcore and an outer, fluid subcore.

coulee. A steep-walled, abandoned channel that glacial meltwater passed through.

country rock. The rock native to an area.

crater. A bowl-shaped, rimmed feature forming the summit of a volcano.

craton. The stable interior of a continent, constructed largely of Precambrian-age rock.

crossbedding. A sequence of inclined sedimentary layers formed by either wind or flowing water.

crust. The outermost shell of Earth, which varies in thickness from 6 to 25 miles.

dike. A tabular intrusion of igneous rock that cuts across the structure of older rock.

dip. The maximum angle that a bedding plane or fault plane makes with the horizontal.

dissolution. The dissolving of rock through chemical alteration.

dolomite. A variety of limestone rich in magnesium carbonate.

drift. A general term for all rock material transported and deposited by a glacier.

earthquake. Ground displacement caused by the sudden release of built-up stress in the crust of the Earth.

epicenter. The point on the Earth's surface that lies directly above an earthquake's focus.

epicontinental. Pertaining to something situated in the interior of a continent, such as an epicontinental sea.

erosion. The removal of soil and weathered rock by wind, water, ice, or gravity.

erosion surface. A generally flat surface planed by erosion.

extrusive rock. Igneous rock formed from magma erupted onto Earth's surface.

fall line. A contact point between upland and lowland usually characterized by waterfalls.

fault. A break in a rock along which there has been movement of one side relative to the adjacent side.

fault block. A mass of rock that has moved along one side of a fault.

fissure vent. A crack or fissure that forms where a volcanic conduit meets the surface of the Earth.

fold. A bend in rock strata or bedding planes.

foraminifera. Very small marine organisms that form calcareous shells, which are useful in the age dating of sedimentary rock.

formation. The basic cartographic rock unit in geologic mapping. A formation has easily recognizable boundaries that can be traced in the field.

fossil. The remains or evidence of preexisting life.

fumarole. A volcanic vent from which gases and vapors are expelled.

geomorphology. The study of the classification, description, origin, and development of landforms.

geyser. A hot spring that periodically erupts volumes of hot water and steam.

glacial polish. A smoothed bedrock surface caused by glacial abrasion.

glaciation. The formation of glaciers and the effects they have on a landscape.

glacier. A mass of ice that moves outward in all directions due to the stress of its own weight.

gneiss. A coarse-grained metamorphic rock characterized by a banded texture formed by alternating zones of lighter and darker minerals.

Gondwanaland. The continent formed during the early Paleozoic era in the southern hemisphere by the amalgamation of several preexisting microcontinents. The supercontinent of Pangaea formed by the joining of Gondwanaland and Laurasia.

graben. An elongate trough bounded on both sides by faults that dip toward the interior of the trough.

granite. A coarse-grained, light-colored, acidic, intrusive rock rich in quartz and potassium feldspar.

greenstone. A compact, metamorphosed igneous rock that owes its dark green color to the presence of chlorite or epidote minerals.

Grenvillia. One of several ancient microcontinents that collided during the late Proterozoic eon to form the core of North America.

gypsum. A sedimentary rock formed of calcium sulfate through the evaporation of water.

hogback. A ridge with a distinct, sharp summit and steep slopes of equal inclination.

horst. An elongate block of rock that has been uplifted relative to the fault blocks on either side of it. It is bounded by normal faults on its long sides.

hot spot. A volcanic center believed to be the surface expression of a rising plume of hot mantle material.

hydrocarbon. Any gaseous, liquid, or solid compound consisting of carbon and hydrogen, crude oil being a common example.

hydrothermal. Pertaining to the action and products of subsurface accumulations of hot water.

Iapetus Ocean. Considered the predecessor to the present-day Atlantic Ocean, this ocean was destroyed by the amalgamation of northern hemisphere microcontinents during the formation of the landmass of Laurasia.

ice field. An extensive mass of ice distributed across a mountain region and covering all but the highest peaks and ridges.

igneous. A class of rock formed by the crystallization of magma.

ignimbrite. Rock formed by the solidification of a pyroclastic flow.

intrusion. A mass of igneous rock that intruded preexisting rock.

island arc. A line of volcanic islands formed over a subduction zone. The Japanese and Aleutian chains of islands are excellent examples.

karst. Topography formed on and within a carbonate rock foundation and characterized by caves, sinkholes, and other dissolution features.

kimberlite. A dark igneous rock that is often a host rock for diamonds.

lahar. A mixture of mud, rock, wood debris, and water that flows down the slopes of a volcano. Also the deposit of such a flow.

land bridge. A terrestrial connection between continents that permits the migration of organisms.

Laurasia. An ancient landmass formed in the northern hemisphere by the amalgamation of several preexisting microcontinents. The supercontinent Pangaea was formed by the joining of Laurasia and Gondwanaland.

Laurentia. A former microcontinent that today forms the geologic heart of North America. Also known as proto–North America.

lava. Molten rock that has erupted onto Earth's surface.

lava dome. A dome-shaped mountain formed by the extrusion of many layers of very fluid lava.

limestone. A sedimentary rock composed chiefly of calcium carbonate. Typically forms in lakes and warm, shallow seas.

lithosphere. The rigid, outermost layer of Earth, composed of the crust and the uppermost portion of the mantle.

magma. Molten or partially molten rock found in the interior of Earth. Called lava when it erupts onto the surface.

mantle. The zone of the Earth that lies below the crust and above the core.

marble. The metamorphic form of a carbonate rock.

marlstone. An earthy or impure form of calcium carbonate mixed with clay.

massive. Said of rock with homogenous texture and lacking layering, fractures, or other discontinuities.

mass wasting. The dislodgement and downslope movement of rock and soil material due to gravity.

meander. One of a series of sinuous curves that define the course of a river.

mesa. An isolated, flat-topped, steep-sloped hill with a summit more extensive than that of a butte.

metamorphic. A class of rock that has been changed in appearance due to exposure to excessively high temperatures and pressures.

metasedimentary. A sedimentary rock that shows evidence of having been altered by metamorphism.

meteorite. A celestial mass of rock that fell to Earth.

microcontinent. A portion of continental lithosphere that broke away from a larger continent.

Mid-Atlantic Ridge. A divergent tectonic plate boundary in the middle of the Atlantic Ocean. It separates the North American Plate from the Eurasian Plate and the South American Plate from the African Plate.

monadnock. A conspicuous hill that stands above the general level of the countryside.

monolith. A large, upstanding, and generally unfractured mass of rock.

moraine. A ridge of drift that accumulated along the front and sides of a glacier.

mudflow. A very fluid, flowing mass of fine-grained sediment.

mudstone. A massive, bedded, indurated mud having the texture and composition of shale but lacking the fine layering that is characteristic of shale.

nonconformity. An unconformity in which sedimentary rocks were deposited on older, eroded igneous or metamorphic rocks.

normal fault. A break in rock, formed by forces of tectonic extension, in which one fault block moves down relative to the block on the other side of the fault.

nunatak. An isolated mass of rock that projects above, and is completely surrounded by, glacial ice.

obsidian. A dark volcanic glass that fractures in a conchoidal pattern.

oil seep. The emergence of migrating, liquid petroleum at the surface of the Earth.

orogeny. The general tectonic process in which mountains are formed.

outcrop. The part of a formation or geologic structure that is exposed at the surface of the Earth.

outwash. Sand and gravel removed from a glacier by meltwater streams and deposited at or beyond the margin of an active glacier.

paleontology. The study of life of the geologic past, as based on fossil plants and animals.

Pangaea. The supercontinent that formed during the late Paleozoic era through the merger of Gondwanaland and Laurasia.

Panthalassa. The ancient ocean that surrounded the supercontinent Pangaea.

pebble. A generally rounded stone between 0.16 and 2.5 inches (4 and 64 millimeters) in diameter.

peneplain. A low-relief, nearly featureless land surface of considerable area.

Piedmont Province. A plateau composed of igneous and metamorphic terranes constructed over an interval of millions of years as the supercontinent Pangaea was being assembled. It is traceable from New Jersey to central Alabama, extends inland to the Blue Ridge Mountains, and reaches a maximum width of 300 miles in North Carolina.

plate. A rigid segment of Earth's lithosphere that moves laterally across Earth's surface and interacts with other similar plates along zones of earthquake and volcanic activity.

plate tectonics. The theory which posits that most large-scale features of Earth form through the relative movement and interaction of the rigid plates composing the lithosphere.

playa. A shallow, ephemeral lake located in an arid or semi-arid region.

plunge pool. A deep, circular lake at the foot of a former waterfall. The force of the falling water scoured out the basin of the lake.

pluton. An igneous intrusion with deep roots in the crust.

pothole. A cylindrical hollow formed in the rocky bed of a river by the grinding action of particles of rock.

precipitate. Name given to a solid that forms from a liquid solution, generally water. For example, limestone.

pumice. A light-colored, vesicular volcanic rock that is often buoyant enough to float on water.

pyroclastic. Fragmented rock material formed by the explosive eruption of a volcano.

quartz. A common rock-forming mineral composed of silicon and oxygen. Typically clear but can come in many colors due to the inclusion of certain elements.

quartzite. A quartz-rich sandstone that has been metamorphosed.

red beds. Sedimentary strata that are predominantly red due to the presence of ferric oxide, an iron oxide.

reef. A moundlike structure built principally of and by calcareous marine organisms.

reservoir rock. A porous and permeable rock that contains oil or gas.

rift. A long and narrow trough bounded by normal faults that completely penetrate the lithosphere. A rift forms as a continent is pulled apart by tectonic extension.

ripple mark. An undulatory rock surface, composed of small-scale ridges and hollows, formed as wind or flowing water moved sediment around before it hardened into rock.

Rodinia. The supercontinent that formed toward the end of the Proterozoic eon; the predecessor of the supercontinent Pangaea.

sand. A sedimentary particle ranging in size between 0.0025 and 0.08 inch (0.06 and 2 millimeters) in diameter.

sandstone. A sedimentary rock composed of rounded, sand-sized sediment that is held together by a naturally occurring cement.

scarp. A small, clifflike slope of considerable linear extent formed by movement along a fault during an earthquake.

schist. A medium- to coarse-grained metamorphic rock rich in minerals oriented in a manner to produce thin layering.

seamount. A flat-topped or peaked and somewhat isolated volcanic hill that rises several thousand feet or more above the seafloor.

sediment. Unconsolidated, solid particles that originate from the weathering and erosion of rock.

sedimentary. A class of rock formed by the deposition and cementation of sediment.

sedimentation. The deposition of sediment by wind, water, or ice.

shale. A sedimentary rock composed of clay- and silt-sized sediment and having a tendency to break along thin, parallel planes.

shatter cone. A striated, conical rock structure formed by shock waves generated by meteorite impact.

Siberia. An ancient microcontinent of which present-day Russia east of the Ural Mountains is composed.

silica. Silicon dioxide; a chemical compound composed of silicon and oxygen. The chief constituent of quartz.

sill. A tabular, intrusive rock that parallels the bedding of the rock it intruded.

silt. A sedimentary particle ranging in size between 0.00016 and 0.0025 inch (0.004 and 0.06 millimeter) in diameter.

siltstone. A sedimentary rock composed of silt-sized sediment with a texture that is intermediate between sandstone and shale.

sinkhole. A circular topographic depression in an area of karst.

slate. Shale that has been metamorphosed.

source rock. Sedimentary rock containing organic material that under the influence of temperature and time can be transformed into liquid or gaseous hydrocarbons.

speleothem. Any of various types of mineral deposits that form in caves.

stalactite. A speleothem that hangs from the ceiling of a cave and develops as minerals precipitate from dripping water.

stalagmite. A speleothem that grows upward from the floor of a cave and develops as minerals precipitate from water that drips onto the floor.

stock. A relatively small intrusive rock mass having an extent of less than 40 square miles.

strata. Tabular layers of sedimentary rock separated from each other by a distinctive bedding plane.

stratovolcano. A volcano that is constructed of alternating layers of pyroclastic material and beds of lava.

striation. One of many minute, parallel lines scratched onto a rock surface by the grinding action of a glacier.

subduction zone. An elongate, narrow belt along which one tectonic plate descends beneath another, as in the Aleutian Island trench.

supercontinent. A landmass formed periodically by the amalgamation of all the existing continents. See **Pangaea** and **Rodinia**.

supercontinent cycle. The periodic aggregation and bifurcation of a supercontinent, on the order of 300 to 500 million years, by way of plate tectonic processes.

syncline. A concave-upward fold in layered rock generally caused by compressive tectonic forces.

Taconica. An ancient island arc, similar to Avalonia, that was added to the ancient core of North America during the early Paleozoic era.

Taconic orogeny. The first of the three Paleozoic-age phases of mountain building that were instrumental in the development of the Appalachian Mountains.

talus. An assemblage of angular rock fragments lying at the base of a cliff or steep slope.

tectonics. Pertaining to the forces involved in the development of the broad architecture of Earth's crust, such as ocean basins, mountains, folds, and faults. See also **plate tectonics**.

terrain. A region of the Earth that is considered a physical feature, such as the Great Plains.

terrane. A body of rock bounded by faults and characterized by a geologic history that differs from adjacent terranes.

Tethys Sea. The ancient sea that existed during Mesozoic time between the protocontinents Gondwanaland and Laurasia.

thermal plume. A vertical column of molten mantle that rises beneath a continent or an ocean. At the surface of the Earth it manifests as a hot spot.

Tippecanoe Sea. An extension, or arm, of the Iapetus Ocean that covered much of what is today North America during the Ordovician and Silurian periods.

trackway. A continuous series of tracks formed by a single organism.

tsunami. A long-wavelength sea wave produced by a large-scale disturbance of the ocean floor, such as a submarine earthquake.

tuff. Consolidated volcanic ash.

unconformity. A gap in the geologic record, generally within a series of sedimentary rocks, in which a sequence of strata is missing either because of erosion or because no sediment was being deposited at that time.

Valley and Ridge Province. That region of the Appalachian Mountains that consists of elongate parallel ridges and valleys that are underlain by thick sequences of folded sedimentary rock of Paleozoic age.

vent. An opening at the surface of the Earth through which volcanic material is channeled.

volcanic. Pertaining to the activities, structures, or rock types associated with a volcano.

volcanic glass. A natural glass produced by the rapid cooling of lava. A common example is obsidian.

water gap. A deep pass in a mountain ridge eroded by flowing water.

weathering. The physical and chemical breakdown of rocks at Earth's surface due to exposure to the atmosphere.

Western Interior Seaway. A sea that extended south from the Arctic Ocean to the Gulf of Mexico and 600 miles from the embryonic Rockies east to the Appalachians, engulfing the central portion of the North America during the Cretaceous period.

wind gap. A former water gap that is no longer associated with flowing water.

zircon. A mineral composed of zirconium silicate and commonly employed in the absolute age dating of rocks because of its resistance to weathering and erosion.

References

Introduction

Dean, D. R. 1992. *James Hutton and the history of geology*. Ithaca, NY: Cornell Univ. Press.

Hallam, A. 1989. *Great geological controversies*. New York, NY: Oxford Univ. Press.

Jackson, P. W. 2006. *The chronologers' quest: The search for the age of the Earth*. Cambridge, UK: Cambridge Univ. Press.

Lawrence, D. M. 2002. *Upheaval from the abyss: Ocean floor mapping and the earth science revolution*. Piscataway, NJ: Rutgers Univ. Press.

McCoy, R. M. 2006. *Ending in ice: The revolutionary idea and tragic expedition of Alfred Wegener*. New York, NY: Oxford Univ. Press.

Winchester, S. 2002. *The map that changed the world: William Smith and the birth of modern geology*. New York, NY: Perennial Harper Collins Publishers.

1. Wetumpka Crater, Alabama

Auburn Astronomical Society. Field Trips. *Wetumpka Meteor Crater*. www.auburnastro.org/wetu.htm.

King, D. T., Jr. *Wetumpka impact crater page*. Geology Dept., Auburn Univ. http://www.auburn.edu/~kingdat/wetumpkawebpage3.htm.

King, D. T., Jr., and L. W. Petruny. 2008. *Wetumpka impact crater guidebook*. Auburn, AL: Parsimony Press.

Neathery, T. L., et al., eds. 1997. *The Wetumpka impact structure and related features*. Alabama Geological Society guidebook 34c.

2. Exit Glacier, Alaska

Follows, D. S. 1990. Kenai Fjords National Park. In *Geology of national parks*, eds. A. G. Harris, et al., p. 375–83. Dubuque, IA: Kendall Hunt Pub. Co.

Miller, D. W. 1986. The Kenai Fjords, where mountains meet the sea. Alaska's Gulf Coast. *Alaska Geographic* 13 (1):63–81.

Musolf, G., et al. 2005. *Glacial geology: Ice on the land*. www.iceagetrail.org/books-and-links.

National Park Service. Kenai Fjords National Park. *Exit Glacier*. www.nps.gov/kefj/planyourvisit/exit-glacier.htm.

3. Antelope Canyon, Arizona

Orndorff, R. L., and D. G. Futey. 2007. *Landforms of southern Utah: A photographic exploration*. Missoula, MT: Mountain Press Pub. Co.

Orndorff, R. L., et al. 2006. *Geology underfoot in southern Utah*. Missoula, MT: Mountain Press Pub. Co.

Photo travelers' guide to slot canyons of the Southwest. 1994. Los Angeles, CA: Photo Traveler Publications.

4. Meteor Crater, Arizona

Kieffer, S. W. 1971. Shock metamorphism of the Coconino sandstone at Meteor Crater, Arizona. *Journal of Geophysical Research* 76 (23):5449–73.

Levy, D. H. 2000. *Shoemaker: The man who made an impact*. Princeton, NJ: Princeton Univ. Press.

Norton, O. R. 1998. *Rocks from space: Meteorites and meteorite hunters*. Missoula, MT: Mountain Press Pub. Co.

Smith, D. 1996. *My meteor crater story*. Winslow, AZ: Meteor Crater Enterprises.

5. Monument Valley, Arizona

Baker, A. A. 1936. *Geology of the Monument Valley–Navajo Mountain region, San Juan County, Utah*. U.S. Geological Survey bulletin 865.

Chenoweth, W. L. 2003. Geology of Monument Valley Navajo Tribal Park, Utah-Arizona. In *Geology of Utah's parks and monuments*, Utah Geological Assoc. publication 28, eds. D. A. Sprinkel, et al., 529–33.

Orndorff, R. L., et al. 2006. *Geology underfoot in southern Utah*. Missoula, MT: Mountain Press Pub. Co.

Railsback, B. Department of Geology, Univ. of Georgia. *A virtual field trip to Monument Valley*. www.gly.uga.edu/railsback/VFT/VFTMonumentValley.html.

6. Prairie Creek Pipe, Arkansas

Henson, P. 1940. Arkansas' Diamond Field. *Gems and Geology* 3 (7):109–12.

Howard, D. L. 1951. Diamond Mines in Arkansas. *Lapidary Journal* 5 (4):253–54.

Worthington, G. W. 2003. *A thorough and accurate history of genuine diamonds in Arkansas.* Murfreesboro, AR: M.A.P./Mid-America Prospecting.

———. 2007. *Genuine diamonds found in Arkansas.* Murfreesboro, AR: M.A.P./Mid-America Prospecting.

7. Wallace Creek, California

Hough, S. E. 2004. *Finding fault in California: An earthquake tourist's guide.* Missoula, MT: Mountain Press Pub. Co.

Sharp, R. P., and A. F. Glazner. 1993. *Geology underfoot in southern California.* Missoula, MT: Mountain Press Pub. Co.

Sieh, K., and R. E. Wallace. 1987. The San Andreas fault at Wallace Creek, San Luis Obispo County, California. In *Cordilleran Section of the Geological Society of America,* centennial field guide, vol. 1, 233–38.

Wallace, R. E. 1990. *The San Andreas fault system, California.* U.S. Geological Survey professional paper 1515.

8. Racetrack Playa, California

Clark, B. 2005. *Death Valley: The story behind the scenery.* Las Vegas, NV: KC Publications Inc.

Messina, P., and P. Stoffer. 2000. Terrain analysis of the Racetrack basin and the sliding rocks of Death Valley. *Geomorphology* 35:253–65.

Sharp, R. P., and D. L. Carey. 1976. Sliding stones, Racetrack Playa, California. *Geological Society of America Bulletin* 87:1704–17.

Sharp, R. P., and A. F. Glazner. 1997. *Geology underfoot in Death Valley and Owens Valley.* Missoula, MT: Mountain Press Pub. Co.

9. Devils Postpile, California

Beard, C. N. 1959. Quantitative study of columnar jointing. *Geological Society of America Bulletin* 70:380–82.

Hartesveldt, R. J. 1952. *The Devil Postpile National Monument.* www.yosemite.ca.us/library/devil_postpile/.

Huber, N. K., and W. W. Eckhardt. 2002. *The story of Devils Postpile.* Three Rivers, CA: Sequoia Natural History Assoc.

Sharp, R. P., and A. F. Glazner. 1997. *Geology underfoot in Death Valley and Owens Valley.* Missoula, MT: Mountain Press Pub. Co.

10. Rancho La Brea, California

Garcia, F. A., and D. S. Miller. 1998. *Discovering fossils: How to find and identify remains of the prehistoric past.* Mechanicsburg, PA: Stackpole Books.

Lange, I. M. 2002. *Ice age mammals of North America: A guide to the big, the hairy, and the bizarre.* Missoula, MT: Mountain Press Pub. Co.

Mestel, R. 1993. Saber-toothed tales. *Discover Magazine* (April). www.discovermagazine.com/1993/apr/sabertoothedtale202.

Reynolds, R. L. 1985. Domestic dog associated with human remains at Rancho La Brea. *Bulletin of the Southern California Academy of Sciences* 84 (2):76–85.

11. El Capitan, California

Alt, D. D., and D. W. Hyndman. 2000. *Roadside geology of northern and central California.* Missoula, MT: Mountain Press Pub. Co.

Huber, N. K. 1989. *The geologic story of Yosemite National Park.* Yosemite National Park, CA: Yosemite Assoc.

Palmer, D. F. 1990. Yosemite National Park. In *Geology of national parks,* eds. A. G. Harris, et al., 298–313. Dubuque, IA: Kendall Hunt Pub. Co.

Schaffer, J. P. 1977. Pleistocene Lake Yosemite and the Wisconsin glaciation of Yosemite Valley. *California Geology* 30 (11):243–48.

12. Boulder Flatirons, Colorado

Chronic, H., and F. Williams. 2005. *Roadside geology of Colorado.* Missoula, MT: Mountain Press Pub. Co.

Johnson, K. R., and R. G. Raynolds. 2003. *Ancient Denvers: Scenes from the past 300 million years of the Colorado Front Range.* Denver, CO: Denver Museum of Nature and Science.

Roach, G. 2008. *Flatiron classics: Easy rock climbs above Boulder.* Colorado Mountain Club guidebooks. Seattle, WA: Mountaineers Books.

Runnells, D. D. 1980. *Boulder, a sight to behold: A guidebook.* Chicago, IL: Johnson Pub. Co.

13. Interstate 70 Roadcut, Colorado

Fay, R. O. 1989. *Geology of the Arbuckle Mountains along Interstate Highway 35, Oklahoma.* Oklahoma Geological Survey guidebook 26.

Ham, W. E., et al. 1969. *Regional geology of the Arbuckle Mountains, Oklahoma.* Oklahoma Geological Survey guidebook 17.

LeRoy, L. W., and R. J. Weimer. 1971. *Geology of the Interstate 70 road cut, Jefferson County, Colorado.* Professional contributions of the Colorado School of Mines 7.

Weimer, R. J., and L. W. LeRoy. 1987. Paleozoic-Mesozoic section: Red Rocks Park, I-70 road cut, and Rooney Road, Morrison area, Jefferson County, Colorado. In *Rocky Mountain Section of the Geological Society of America,* centennial field guide, vol. 2, 315–19.

14. Florissant Fossil Beds, Colorado

Cook, T., and L. Abbott. 2011. Florissant Fossil Beds: An Eocene time capsule. *Earth* 56 (7):48–52.

Manwell, R. D. 1955. An insect Pompeii. *Scientific Monthly* 80 (6):356–61.

Meyer, H. W., and L. Weber. 1995. Florissant Fossil Beds National Monument: Preservation of an ancient ecosystem. *Rocks and Minerals* 70:232–41.

Saenger, W. 1982. *Florissant Fossil Beds National Monument: Window to the past*. Estes Park, CO: Rocky Mountain Nature Assoc., Inc.

15. Dinosaur Trackway, Connecticut

Benton, M. J. 1996. *The Penguin historical atlas of the dinosaurs*. New York, NY: Penguin USA.

Coleman, M. E. 2005. *The geological history of Connecticut's bedrock*. State Geological and Natural History Survey of Connecticut, special publications 2.

McHone, J. G. 2004. *Connecticut in the Mesozoic world*. Connecticut Dept. of Environmental Protection.

Rodgers, J. 1980. The Geological History of Connecticut. *Discovery* 15 (1):3–25.

16. Wilmington Blue Rocks, Delaware

Delaware Geological Survey. *Exploring the Wilmington Complex (the Wilmington blue rocks): A geoadventure in the Delaware Piedmont*. www.dgs.udel.edu/delaware-geology/exploring-wilmington-blue-rocks-geoadventure-delaware-piedmont.

McPhee, J. 1983. *In suspect terrain*. New York, NY: Farrar, Straus, and Giroux.

Plank, M. O., and W. S. Schenck. 1998. *Delaware Piedmont geology*. Delaware Geological Survey special publication 20.

Plank, M. O., et al. 2000. *Bedrock geology of the Piedmont of Delaware and adjacent Pennsylvania*. Delaware Geological Survey report of investigations 59.

17. Devil's Millhopper, Florida

Bryan, J. R., et al. 2008. *Roadside geology of Florida*. Missoula, MT: Mountain Press Pub. Co.

Gupta, S. 2010. Sinkholes: Florida grapples with the wonders of the not-so-deep. *Earth* 55 (8):44–53.

Scott, T. M. 1986. Devil's Mill hopper, Alachua County, Florida. In *Southeastern Section of the Geological Society of America*, centennial field guide, vol. 6, 335–37.

White, W. A. 1970. *The geomorphology of the Florida peninsula*. Florida Geological Survey bulletin 51.

18. Stone Mountain, Georgia

Grant, W. H. 1986. Structural and petrologic features of the Stone Mountain granite pluton, Georgia. In *Southeastern Section of the Geological Society of America*, centennial field guide, vol. 6, 285–90.

Herrmann, L. A. 1954. *Geology of the Stone Mountain–Lithonia district, Georgia*. Geological Survey of Georgia bulletin 61.

Hopson, C. A. 1958. Exfoliation and weathering at Stone Mountain, Georgia. *Georgia Mineralogical Newsletter* 3:65–79.

Hudson, P. S., and L. P. Mirza. 2011. *Atlanta's Stone Mountain: A multicultural history*. Charleston, SC: The History Press.

19. Kilauea Volcano, Hawaii

Fisher, R. V., et al. 1997. *Volcanoes: Crucibles of change*. Princeton, NJ: Princeton Univ. Press.

Hazlett, R. W. 1987. Kilauea caldera and adjoining volcanic rift zones. In *Cordilleran Section of the Geological Society of America*, centennial field guide, vol. 1, 15–20.

Kornprobst, J., and C. Laverne. 2006. *Living mountains: How and why volcanoes erupt*. Missoula, MT: Mountain Press Pub. Co.

U.S. Geological Survey Hawaiian Volcano Observatory. *Recent Kilauea status reports, updates, and information releases*. http://volcanoes.usgs.gov/hvo/activity/kilaueastatus.php.

20. Borah Peak, Idaho

Alt, D. D., and D. W. Hyndman. 1989. *Roadside geology of Idaho*. Missoula, MT: Mountain Press Pub. Co.

Crone, A. J. 1987. Surface faulting associated with the 1983 Borah Peak earthquake at Doublespring Pass Road, east-central Idaho. In *Rocky Mountain Section of the Geological Society of America*, centennial field guide, vol. 2, 95–98.

Crone, A. J., and M. N. Machette. 1984. Surface faulting accompanying the Borah Peak earthquake, central Idaho. *Geology* 12 (11):664–67.

Earthquake . . . . 1984. Available from the Arco Advertiser Inc., 146 South Front Street, Arco, ID, 83213.

21. Menan Buttes, Idaho

Creighton, D. N. 1987. Menan Buttes, southeastern Idaho. In *Rocky Mountain Section of the Geological Society of America*, centennial field guide, vol. 2, 109–11.

Hamilton, W., and W. B. Myers. 1963. *Menan Buttes, cones of glassy basalt tuff in the Snake River Plain, Idaho*. U.S. Geological Survey professional paper 450E.

Sterns, H. T., et al. 1938. *Geology and groundwater resources of the Snake River Plain in southeastern Idaho*. U.S. Geological Survey water supply paper 774.

Wood, C. A., and J. Kienle. 1992. *Volcanoes of North America: United States and Canada*. Cambridge, UK: Cambridge Univ. Press.

22. Great Rift, Idaho

Alt, D. D., and D. W. Hyndman. 1989. *Roadside geology of Idaho*. Missoula, MT: Mountain Press Pub. Co.

Kuntz, M. A., et al. 1987. Geology of the Craters of the Moon lava field, Idaho. In *Rocky Mountain Section of the Geological Society of America*, centennial field guide, vol. 2, 123–26.

Owen, D. E. 2008. *Geology of Craters of the Moon*. Arco, ID: Craters of the Moon Natural History Assoc.

Walz, J. H. 1993. *Hiking the Great Rift*. www.idahoptv.org/outdoors/shows/path/rift1.html.

23. Valmeyer Anticline, Illinois

Norby, R. D. 1987. Valmeyer anticline of Monroe County, Illinois. In *North-Central Section of the Geological Society of America*, centennial field guide, vol. 3, 237–40.

Reinertsen, D. L. 1981. *A guide to the geology of the Waterloo-Valmeyer area*. Geological science field trip guide leaflet 1981-D. Illinois State Geological Survey.

Swann, D. H. Illinois Oil and Gas Assoc. *A summary geologic history of the Illinois Basin*. http://www.ioga.com/Special/Geohist.htm.

Treworgy, J. D. 1981. *Structural features in Illinois—a compendium*. Illinois State Geological Survey circular 519.

24. Hanging Rock Klint, Indiana

Camp, M. J., and Graham T. R. 1999. *Roadside geology of Indiana*. Missoula, MT: Mountain Press Pub. Co.

Shaver, R. H. 1987. The Silurian reefs near Wabash, Indiana. In *North-Central Section of the Geological Society of America*, centennial field guide, vol. 3, 333–36.

Shrock, R. R. 1929. The klintar of the upper Wabash Valley in northern Indiana. *Journal of Geology* 37 (1):17–29.

Textoris, D. A., and A. V. Caroz. 1964. Petrography and evolution of Niagaran (Silurian) reefs, Indiana. *American Association of Petroleum Geologists Bulletin* 48 (4):397–426.

25. Fort Dodge Gypsum, Iowa

Anderson, R. R. Iowa Department of Natural Resources, Iowa Geological and Water Survey. *Fort Dodge gypsum: A salt from Iowa's Jurassic sea*. http://www.igsb.uiowa.edu/Browse/ftdodge/ftdodge.htm.

Anderson, R. R., and R. M. McKay. 1999. *The geology of the Jurassic Fort Dodge Formation, Webster County, Iowa*. Geological Society of Iowa guidebook 67.

Geological Society of Iowa. 1976. *Geologic points of interest in the Fort Dodge area*. Geological Society of Iowa guidebook 28.

Troeger, J. 1983. *From rift to drift: Iowa's story in stone*. Ames, IA: Iowa State Univ. Press.

26. Monument Rocks, Kansas

Buchanan, R. C., and J. R. McCauley. 2010. *Roadside Kansas: A traveler's guide to its geology and landmarks*. Lawrence, KS: Univ. Press of Kansas for the Kansas Geological Survey.

GeoKansas. *Field trip highlights—northwestern Kansas*. www.kgs.ku.edu/Extension/fieldtrips/ESW2002.html.

McCauley, J. R., et al. 1997. *Fossil collecting in the Cretaceous Niobrara chalk*. Kansas Geological Survey open-file report 97-62.

Rogers, K. L. 1999. *The Sternberg fossil hunters: A dinosaur dynasty*. Missoula, MT: Mountain Press Pub. Co.

27. Ohio Black Shale, Kentucky

Kepferle, R. C. 1986. Devonian and Mississippian black shales of Kentucky. In *Southeastern Section of the Geological Society of America*, centennial field guide, vol. 6, 31–36.

Kepferle, R. C. 2001. *Devonian system*. http://pubs.usgs.gov/prof/p1151h/devonian.html. U.S. Geological Survey contributions to the geology of Kentucky.

Provo, L. J., et al. 1978. Division of black Ohio shale in eastern Kentucky. *American Association of Petroleum Geologists Bulletin* 62 (9):1703–13.

Schopf, J. M., and J. F. Schwietering. 1970. *The Foerstia zone of the Ohio and Chattanooga shales*. U.S. Geological Survey bulletin 1294-H.

28. Mammoth Cave, Kentucky

Lyons, J. M. 2005. *Mammoth Cave: The story behind the scenery*. Las Vegas, NV: KC Publications Inc.

Palmer, A. N. 1981. *A geological guide to Mammoth Cave National Park*. Teaneck, NJ: Zephyrus Press.

———. 1990. Mammoth Cave National Park. In *Geology of national parks*, eds. A. G. Harris, et al., 145–58. Dubuque, IA: Kendall Hunt Pub. Co.

Watson, P. J. 1974. *The archaeology of the Mammoth Cave area*. Salt Lake City, UT: Academic Press.

29. Four Corners Roadcut, Kentucky

Chesnut, D. R., et al. 1986. "Four Corners": A 3-dimensional panorama of stratigraphy and depositional structures of the Breathitt Formation (Pennsylvanian), Hazard, Kentucky. In *Southeastern Section of the Geological Society of America*, centennial field guide, vol. 6, 47–50.

Cobb, J. C., et al. 1981. *Coal and coal-bearing rocks of eastern Kentucky*. Geological Society of America Coal Division field trip. Kentucky Geological Survey.

Ruppert, L., et al. 2002. *Coal resources of selected coal beds and zones in the northern and central Appalachian Basin*. U.S. Geological Survey fact sheet 004-02. http://pubs.usgs.gov/fs/fs004-02/fs004-02.html.

Seiders, V. M. 1964. *Geology of the Hazard north quadrangle*. U.S. Geological Survey geologic quadrangle map GQ 344.

30. Avery Island, Lousiana

Atwater, G. I., and M. J. Forman. 1959. Nature of growth of southern Louisiana salt domes and its effect on petroleum accumulations. *American Association of Petroleum Geologists Bulletin* 43 (11):2592–622.

Halbouty, M. T. 1981. *Salt domes: Gulf region, United States and Mexico*. Houston, TX: Gulf Pub. Co.

Kupfer, D. H. 1986. Physiography of Louisiana salt domes (the five islands). In *Southeastern Section of the Geological Society of America*, centennial field guide, vol. 6, 431–34.

Murray, G. E. 1961. *Geology of the Atlantic and Gulf coastal province of North America*. New York, NY: Harper's geoscience series.

31. Schoodic Point, Maine

Acadia National Park. 1990. In *Geology of national parks*, eds. A. G. Harris, et al., 247–59. Dubuque, IA: Kendall Hunt Pub. Co.

Butcher, R. D. 1977. *Field guide to Acadia National Park, Maine*. NY: Reader's Digest Press.

Chapman, C. A. 1970. *The geology of Acadia National Park*. Old Greenwich, CT: Chatham Press.

Gilman, R. A., et al. 1988. *The geology of Mount Desert Island: A visitor's guide to the geology of Acadia National Park*. Maine Geological Survey.

32. Calvert Cliffs, Maryland

Fossilguy.com. *Shark fossils found at Calvert Cliffs of Maryland*. http://www.fossilguy.com/sites/calvert/calv_srk.htm.

Gernant, R. E., et al. 1971. *Environmental history of Maryland Miocene*. Maryland Geological Survey guidebook 3.

Glaser, J. D. 1995. *Collecting fossils in Maryland*. Maryland Geological Survey educational series 4.

Vogt, P. R., and R. Eshelman. 1987. Maryland's cliffs of Calvert: A fossiliferous record of mid-Miocene inner shelf and coastal environments. In *Northeastern Section of the Geological Society of America*, centennial field guide, vol. 5, 9–14.

33. Purgatory Chasm, Massachusetts

Pelto, M. S. *Purgatory Chasm State Reservation: How did it form?* http://www.nichols.edu/departments/purgatorychasm/index.htm.

Raymo, C., and M. E. Raymo. 2001. *Written in stone: A geological history of the northeastern United States*. Delmar, NY: Black Dome Press Corp.

Roberts, D. C. 2001. *A field guide to geology: Eastern North America*. Boston, MA: Houghton Mifflin Harcourt.

Skehan, J. W. 2001. *Roadside geology of Massachusetts*. Missoula, MT: Mountain Press Pub. Co.

34. Nonesuch Potholes, Michigan

Dorr, J. A., and D. F. Eschman. 1970. *Geology of Michigan*. Ann Arbor, MI: Univ. of Michigan Press.

Kelly, W. C., and G. K. Nishioka. 1985. Precambrian oil inclusions in late veins and the role of hydrocarbons in copper mineralization at White Pine, Michigan. *Geology* 13 (5):334–37.

Mudrey, M. G., Jr., ed. 1986. *Precambrian petroleum potential, Wisconsin and Michigan*. Geoscience Wisconsin 11. Madison, WI: Univ. of Wisconsin—Extension, Wisconsin Geological and Natural History Survey.

Reed, R. C. 1987. Porcupine Mountain's Wilderness State Park, Michigan. In *North-Central Section of the Geological Society of America*, centennial field guide, vol. 3, 269–72.

35. Quincy Mine, Michigan

Butler, B. S., and W. S. Burbank. 1929. *The copper deposits of Michigan*. U.S. Geological Survey professional paper 144.

Hunt, M. H., and D. Hunt. 2001. *Hunts' guide to Michigan's Upper Peninsula*. Hancock, MI: Midwestern Guides Inc.

Lankton, L. D., and C. K. Hyde. 2003. *Old reliable: An illustrated history of the Quincy Mining Company*. Hancock, MI: Quincy Mine Hoist Assoc.

Murdoch, A. 2001. *Boom copper: The story of the first U.S. mining boom*. Hancock, MI: Quincy Mine Hoist Assoc.

36. Grand River Ledges, Michigan

Dorr, J. A., and D. F. Eschman. 1970. *Geology of Michigan*. Ann Arbor, MI: Univ. of Michigan Press.

Kelly, W. A. 1933. Pennsylvanian stratigraphy near Grand Ledge, Michigan. *Journal of Geology* 41 (1):77–88.

The ledges of Grand Ledge. http://wsharing.com/WSphotosLedges1.htm.

Milstein, R. L. 1987. The ledges of the Grand River, Michigan. In *North-Central Section of the Geological Society of America*, centennial field guide, vol. 3, 311–14.

37. Sioux Quartzite, Minnesota

Corbett, W. P. 1980. Pipestone: The origin and development of a national monument. *Minnesota History* 47 (fall):83–92.

Morey, G. B., and D. R. Setterholm. 1987. Pipestone National Monument: The Sioux quartzite—an early Proterozoic braided stream deposit, southwestern Minnesota. In *North-Central Section of the Geological Society of America*, centennial field guide, vol. 3, 73–76.

Murray, R. A. 1965. *A history of Pipestone National Monument, Minnesota*. Pipestone, MN: Pipestone Indian Shrine Assoc.

Pipestone County Historical Society. *A brief history of Pipestone, Minnesota*. http://www.pipestoneminnesota.com/museum/history2.htm.

38. Thomson Dikes, Minnesota

Jirsa, M. A., and G. B. Morey. 1987. Jay Cooke State Park and Grandview areas: Evidence for a major early Proterozoic–middle Proterozoic unconformity in Minnesota. In *North-Central Section of the Geological Society of America*, centennial field guide, vol. 3, 67–72.

Morey, G. B., and R. W. Ojakangas. 1970. *Sedimentology of the middle Precambrian Thomson Formation, east-central Minnesota*. Minnesota Geological Survey report of investigations 13.

Ojakangas, R. W. 2009. *Roadside geology of Minnesota*. Missoula, MT: Mountain Press Pub. Co.

Sansome, C. J. 1990. *Minnesota underfoot: A field guide to Minnesota's geology*. Minneapolis, MN: Voyageur Press (Quayside).

39. Soudan Mine, Minnesota

Ojakangas, R. W. 2009. *Roadside geology of Minnesota*. Missoula, MT: Mountain Press Pub. Co.

Ojakangas, R. W., and C. L. Matsch. 1982. *Minnesota's geology*. Minneapolis, MN: Univ. of Minnesota Press.

Sims, P. K., and G. B. Morey. 1966. *Geologic sketch of the Tower-Soudan State Park*. Minnesota Geological Survey educational series 3.

Walker, D. A. 2004. *Iron frontier: The discovery and early development of Minnesota's three ranges*. St. Paul, MN: Minnesota Historical Society Press.

40. Petrified Forest, Mississippi

Blackwell, W. H., et al. 1981. The structural and phytogeographic affinities of some silicified wood from the mid-Tertiary of west-central Mississippi. In *Geobotany II*, ed. R. C. Romans. New York, NY: Plenum Press.

Hickey, L. J. 2010. *The forest primeval: The geologic history of wood and petrified forests*. New Haven, CT: The Yale Peabody Museum.

Schabilion, S. 1976. *Mississippi Petrified Forest: A place of fascination*. Flora, MS: MPF Publishers.

———. 1996. *Out of the past: Mississippi Petrified Forest: A scrapbook of time*. Louisville, KY: Southern Pub. Co. Inc.

41. Elephant Rocks, Missouri

Graves, H. B. 1938. The Pre-Cambrian structure of Missouri. *Transactions of the Missouri Academy of Science* 29:111–61.

Kisvarsanyi, E. B., and A. W. Hebrank. 1987. Elephant Rocks: A granite tor in Precambrian Graniteville granite, the St. Francois Mountains, Missouri. In *North-Central Section of the Geological Society of America*, centennial field guide, vol. 3, 159–60.

Tolman, C., and F. Robertson. 1969. *Exposed Precambrian rocks in southeast Missouri*. Missouri Geological Survey and Water Resources report of investigations 44.

Watkins, C. 2006. Conor Watkins' Ozark Mountain Experience, Article 6 and 7. *The St. Francois Mountains—Missouri's hard rock core*. http://www.rollanet.org/~conorw/cwome/article6&7.htm.

42. Grassy Mountain Nonconformity, Missouri

Kisvarsanyi, E. B., and A. W. Hebrank. 1987. Roadcuts in the St. François Mountains, Missouri: Basalt-dike swarm in granite, Precambrian-Paleozoic nonconformity and a Lamotte channel-fill deposit. In *North-Central Section of the Geological Society of America*, centennial field guide, vol. 3, 161–64.

Spencer, C. G. 2011. *Roadside geology of Missouri*. Missoula, MT: Mountain Press Pub. Co.

Tolman, C., and F. Robertson. 1969. *Exposed Precambrian rocks in southeast Missouri*. Report of investigations 44. Contribution to Precambrian geology 1. Missouri Geological Survey and Water Resources.

Whittington, A. 2005. *St. François Mountains: Igneous and metamorphic geology field trip*. http://web.missouri.edu/~whittingtona/photos/SEMO2005/index.html.

43. Chief Mountain, Montana

Alt, D. D., and D. W. Hyndman. 1995. *Northwest exposures: A geologic story of the Northwest*. Missoula, MT: Mountain Press Pub. Co.

Beaumont, G. 1978. *Many-storied mountains: The life of Glacier National Park*. National Park Service natural history series.

Mudge, M. R., and R. L. Earhart. 1980. *The Lewis thrust fault and related structures in the disturbed belt, northwestern Montana*. U.S. Geological Survey professional paper 1174.

Waterton-Glacier International Peace Park. 1990. In *Geology of national parks*, eds. A. G. Harris, et al., 277–88. Dubuque, IA: Kendall Hunt Pub. Co.

44. Madison Slide, Montana

Alt, D. D., and D. W. Hyndman. 1986. *Roadside geology of Montana*. Missoula, MT: Mountain Press Pub. Co.

Ryall, A. 1962. The Hebgen Lake, Montana, earthquake of August 17, 1959. *Bulletin of the Seismological Society of America* 52 (2): 235–71.

Witkind, I. J., and M. C. Stickney. 1987. The Hebgen Lake earthquake area, Montana and Wyoming. In *Rocky Mountain Section of the Geological Society of America*, centennial field guide, vol. 2, 89–94.

Witkind, I. J., et al. 1962. Geologic features of the earthquake at Hebgen Lake, Montana, August 17, 1959. *Bulletin of the Seismological Society of America* 52 (2):163–80.

45. Butte Pluton, Montana

Alt, D. D., and D. W. Hyndman. 1986. *Roadside geology of Montana*. Missoula, MT: Mountain Press Pub. Co.

Glasscock, C. B. 1935. *The war of the copper kings*. New York, NY: Grosset and Dunlap.

Malone, M. P. 1981. *The battle for Butte: Mining and politics on the northern frontier, 1864–1906*. Seattle, WA: Univ. of Washington Press.

Zeihen, L. G., et al. 1987. Geology of the Butte mining district, Montana. In *Rocky Mountain Section of the Geological Society of America*, centennial field guide, vol. 2, 57–61.

46. Quad Creek Quartzite, Montana

Bowring, S. A. 1989. 3.96 Ga gneisses from the Slave Province, Northwest Territories, Canada. *Geology* 17 (11):971–75.

James, H. L. 1995. *Geologic and historic guide to the Beartooth Highway, Montana and Wyoming*. Montana Bureau of Mines and Geology special publication 110.

Mueller, P. A., et al. 1987. A study in contrasts: Archean and Quaternary geology of the Beartooth Highway, Montana and Wyoming. In *Rocky Mountain Section of the Geological Society of America*, centennial field guide, vol. 2, 75–78.

———. 1992. 3.96 Ga zircons from an Archean quartzite, Beartooth Mountains, Montana. *Geology* 20 (4):327–30.

47. Ashfall Fossil Beds, Nebraska

Bouc, K. 1994. *The cellars of time: Paleontology and archaeology in Nebraska*. Lincoln, NE: Nebraska Game and Parks Commission.

Maher, H. D., Jr., et al. 2002. *Roadside geology of Nebraska*. Missoula, MT: Mountain Press Pub. Co.

Rose, W. I., et al. 2003. Sizes and shapes of 10-Ma distal fall pyroclasts in the Ogallala Group, Nebraska. *Journal of Geology* 111:115–24.

Voorhies, M. 1992. *Ashfall: Life and death at a Nebraska waterhole ten million years ago*. Univ. of Nebraska State Museum, museum notes 81. http://www.unl.edu/museum/research/vertpaleo/ashfall.html.

48. Scotts Bluff, Nebraska

Hunter, R. E. 1977. Basic types of stratification in small eolian dunes. *Sedimentology* 24 (3):361–87.

Loope, D. B. 1986. Recognizing and utilizing vertebrate tracks in cross section: Cenozoic hoofprints from Nebraska. *Palaios* 1:141–51.

Maher, H. D., Jr., et al. 2002. *Roadside geology of Nebraska*. Missoula, MT: Mountain Press Pub. Co.

Swinehart, J. B., and D. B. Loope. 1987. Late Cenozoic geology along the summit to museum hiking trail, Scotts Bluff National Monument, western Nebraska. In *North-Central Section of the Geological Society of America*, centennial field guide, vol. 3, 13–18.

49. Crow Creek Marlstone, Nebraska

Crandall, D. R. 1952. Origin of Crow Creek member of Pierre shale in central South Dakota. *American Association of Petroleum Geologists Bulletin* 36 (9):1754–65.

Izett, G. A., et al. 1993. The Manson impact structure: $^{40}AR/^{39}AR$ age and its distal impact ejecta in the Pierre shale in southeastern South Dakota. *Science* 262 (5134):729–32.

Katongo, C., et al. 2004. Geochemistry and shock petrography of the Crow Creek Member, South Dakota, USA: Ejecta from the 74-Ma Manson impact structure. *Meteoritics and Planetary Science* 39 (1):31–51.

Maher, H. D., Jr., et al. 2002. *Roadside geology of Nebraska*. Missoula, MT: Mountain Press Pub. Co.

50. Sand Mountain, Nevada

Bagnold, R. A. 1971. *The physics of blown sand and desert dunes*. London, UK: Chapman and Hall.

Greenburg, G. 2008. *A grain of sand: Nature's secret wonder*. McGregor, MN: Voyageur Press.

Sholtz, P., et al. 1997. Sound-producing sand avalanches. *Contemporary Physics* 38 (5):329–42.

Trexler, D. T., and W. N. Melhorn. 1986. Singing and booming sand dunes of California and Nevada. *California Geology* 39 (7):147–52.

51. Great Unconformity, Nevada

Moreno, R. 2000. *Roadside history of Nevada*. Missoula, MT: Mountain Press Pub. Co.

Rea, D. K., et al. 2006. Broad region of no sediment in the southwest Pacific basin. *Geology* 34 (10):873–76.

Rowland, S. M. 1987. Paleozoic stratigraphy of Frenchman Mountain, Clark County, Nevada. In *Cordilleran Section of the Geological Society of America*, centennial field guide, vol. 1, 53–56.

Shelton, J. S. 1966. *Geology illustrated*. San Francisco, CA: W. H. Freeman and Co.

52. Flume Gorge, New Hampshire

Creasy, J. W., and J. P. Fitzgerald. *Bedrock geology of the eastern White Mountain Batholith, North Conway area, New Hampshire*. http://abacus.bates.edu/acad/depts/geology/jcreasy.WM.html.

Raymo, C., and M. E. Raymo. 2001. *Written in stone: A geological history of the northeastern United States*. Delmar, NY: Black Dome Press Corp.

Van Diver, B. B. 1987. *Roadside geology of Vermont and New Hampshire*. Missoula, MT: Mountain Press Pub. Co.

53. Palisades Sill, New Jersey

Itoi, N. G. 2008. *Hudson River Valley*. Jackson, TN: Avalon Travel Publishing, Perseus Books Group.

Puffer, J. H., et al. 1992. *The Palisades Sill and Watchung basalt flows, northern New Jersey and southeastern New York: A geological summary and field guide*. New Jersey Geological Survey open-file report 92-1.

Serrao, J. 1986. *The wild palisades of the Hudson*. Westwood, NJ: Lind Publications.

Walker, K. 1969. *The Palisades Sill, New Jersey: A reinvestigation*. Boulder, CO: Geological Society of America.

54. White Sands, New Mexico

Atkinson, R. 1977. *White Sands: Wind, sand and time*. Globe, AZ: Southwest Parks and Monuments Assoc.

Houk, R., and M. Collier. 1994. *White Sands National Monument*. Globe, AZ: Southwest Parks and Monuments Assoc.

LeMone, D. V. 1987. White Sands National Monument, New Mexico. In *Rocky Mountain Section of the Geological Society of America*, centennial field guide, vol. 2, 451–54.

McKee, E. D, and J. R. Douglass. 1971. *Growth and movement of dunes at White Sands National Monument, New Mexico*. U.S. Geological Survey professional paper 750-D.

55. Carlsbad Caverns, New Mexico

Burnham, B. 2003. *Carlsbad Caverns: America's largest underground chamber*. New York, NY: Rosen Pub. Group.

Chronic, H. 1987. *Roadside geology of New Mexico*. Missoula, MT: Mountain Press Pub. Co.

Greene, E. J. 2006. *Carlsbad Caverns: The story behind the scenery*. Las Vegas, NV: KC Publications Inc.

White, J. L. 1998. *Jim White's own story: The discovery and history of Carlsbad Caverns*. Carlsbad, NM: Carlsbad Caverns Guadalupe Mountains Assoc.

56. Ship Rock, New Mexico

Chronic, H. 1987. *Roadside geology of New Mexico*. Missoula, MT: Mountain Press Pub. Co.

Delaney, P. T. 1987. Shiprock, New Mexico: The vent of a violent volcanic eruption. In *Rocky Mountain Section of the Geological Society of America*, centennial field guide, vol. 2, 411–15.

Hack, J. T. 1942. Sedimentation and volcanism in the Hopi Buttes, Arizona. *Geological Society of America Bulletin* 53 (2):335–72.

Williams, H. 1936. Pliocene volcanoes of the Navajo-Hopi country. *Geological Society of America Bulletin* 47 (1):111–71.

57. Stateline Outcrop, New Mexico

Anderson, R. Y., and D. W. Kirkland. 1966. Intrabasin varve correlation. *Geological Society of America Bulletin* 77 (3):241–56.

———. 1987. Banded Castile evaporites, Delaware Basin, New Mexico. In *Rocky Mountain Section of the Geological Society of America*, centennial field guide, vol. 2, 455–58.

Chronic, H. 1987. *Roadside geology of New Mexico*. Missoula, MT: Mountain Press Pub. Co.

St. John, J. *Gypsum plain*. http://www1.newark.ohio-state.edu/Professional/OSU/Faculty/jstjohn/New-Mexico-Geology/State-Line-outcrop.htm.

58. American Falls, New York

Brett, C. E., and P. E. Calkin. 1987. Niagara Falls and Gorge, New York–Ontario. In *Northeastern Section of the Geological Society of America*, centennial field guide, vol. 5, 97–105.

Fisher, D. W., and C. E. Brett. 1981. The geologic past. In *Colossal cataract: The geologic history of Niagara Falls*, ed. I. H. Tesmer, 16–62. Albany, NY: State Univ. of New York Press.

Philbrick, S. S. 1974. What future for Niagara Falls? *Geological Society of America Bulletin* 85 (1):91–98.

Van Diver, B. B. 1985. *Roadside geology of New York*. Missoula, MT: Mountain Press Pub. Co.

59. Taconic Unconformity, New York

Fisher, D. W. 2006. *The rise and fall of the Taconic Mountains: A geological history of eastern New York*. Delmar, NY: Black Dome Press Corp.

Isachsen, Y. W. 2000. *Geology of New York: A simplified account*. New York State Museum educational leaflet 28.

Raymo, C., and M. E. Raymo. 2001. *Written in stone: A geological history of the northeastern United States*. Delmar, NY: Black Dome Press Corp.

Van Diver, B. B. 1985. *Roadside geology of New York*. Missoula, MT: Mountain Press Pub. Co.

60. Gilboa Forest, New York

Boyer, J. S., Jr. 1993. *Reexamination of* Eospermalopteris eriana *(Dawson) Goldring from the Upper Middle Devonian (=Givetian) flora at Gilboa, New York*. Thesis, John Carroll Univ. http://mysite.verizon.net/james.s.boyer/Boyer-ThesisNoImages.pdf.

Gensel, P. G., and D. Edwards. 2001. *Plants invade the land: Evolutionary and environmental perspectives*. New York, NY: Columbia Univ. Press.

Hernick, L. 2003. *The Gilboa fossils*. Albany, NY: New York State Museum.

Ment, J. 2008. The Gilboa Museum awaits. *Catskill Mountain Foundation Guide Magazine* (September). http://www.catskillmtn.org/guide-magazine/articles/2008-09-the-gilboa-museum-awaits.html.

61. Pilot Mountain, North Carolina

Butler, J. R. 1986. Pilot Mountain, North Carolina. In *Southeastern Section of the Geological Society of America*, centennial field guide, vol. 6, 227–28.

Carpenter, P. A., III. 1989. *A geologic guide to North Carolina's state parks*. North Carolina Geological Survey bulletin 91.

Kesel, R. N. 1974. Inselbergs on the piedmont of Virginia, North Carolina, and South Carolina: Types and characteristics. *Southeastern Geology* 16 (1):1–30.

Stewart, K. G., and M. R. Roberson. 2007. *Exploring the geology of the Carolinas: A field guide to favorite places from Chimney Rock to Charleston*. Chapel Hill, NC: Univ. of North Carolina Press.

62. South Killdeer Mountain, North Dakota

Bluemle, J. 2002. *North Dakota's mountainous areas: The Killdeer Mountains and the Turtle Mountains*. North Dakota Geological Survey, North Dakota notes 15. http://www.dmr.nd.gov/ndgs/ndnotes/ndn15-h.htm.

Forney, G. G. 1977. Medicine Hole, Dunn County, North Dakota. *Windy City Speleonews* 17 (August):68–71.

Hoganson, J. W., and E. C. Murphy. 2003. *Geology of the Lewis and Clark Trail in North Dakota*. Missoula, MT: Mountain Press Pub. Co.

Murphy, E. C., et al. 1993. *The Chadron, Brule and Arikaree formations of North Dakota*. North Dakota Geological Survey report of investigation 96.

63. Hueston Woods, Ohio

Davis, R. A. 1992. *Cincinnati fossils: An elementary guide to the Ordovician rocks and fossils of the Cincinnati, Ohio, region*. Cincinnati Museum of Natural History popular publication series 10.

Feldmann, R. M., and M. Hackathorn, eds. 2005. *Fossils of Ohio*. Ohio Division of Geological Survey bulletin 70.

Shrake, D. L. 2003. *Fossil collecting in Ohio*. Ohio Division of Geological Survey geofacts 17. http://www.dnr.state.oh.us/Portals/10/pdf/GeoFacts/geof17.pdf.

———. 2005. *Ohio trilobites*. Ohio Division of Geological Survey geofacts 5. http://www.dnr.state.oh.us/Portals/10/pdf/GeoFacts/geof05.pdf.

64. Big Rock, Ohio

Camp, M. J. 2006. *Roadside geology of Ohio*. Missoula, MT: Mountain Press Pub. Co.

Hansen, M. C. 1984. Glacial erratics, or "What's the biggest rock in Ohio?" *Ohio Geology Newsletter* (winter):1–5.

Hussey, J. 1878. Geology of Shelby County. In *Report of the Geological Survey of Ohio vol. 3: Geology and paleontology* 448–67. Columbus, OH: Nevins and Myers State Printers.

Peters, R. L., and G. Faure. 1972. Age determination of a glacial erratic in Columbus, Ohio. *Ohio Journal of Science* 72 (2):87–90.

65. Kelleys Island, Ohio

Camp, M. J. 2006. *Roadside geology of Ohio*. Missoula, MT: Mountain Press Pub. Co.

Feldmann, R. M., and T. W. Bjerstedt. 1987. Kelleys Island: Giant glacial grooves and Devonian shelf carbonates in north-central Ohio. In *North-Central Section of the Geological Society of America*, centennial field guide, vol. 3, 395–98.

Hansen, M. C. 1988. Glacial grooves: "Rock-scorings of the great ice invasions": Revisited. *Ohio Geology Newsletter* (spring):1–5.

Snow, R. S., et al. 1991. A field guide: The Kelleys Island glacial grooves, subglacial erosion features on the Marblehead Peninsula, carbonate petrology, and associated paleontology. *Ohio Journal of Science* 91 (1):16–26.

66. Interstate 35 Roadcut, Oklahoma

Alberstadt, L. P. 1973. *Articulate brachiopods of the Viola Formation (Ordovician) in the Arbuckle Mountains, Oklahoma*. Oklahoma Geological Survey bulletin 117.

Brown, R. L. 2002. The birth of the seismic reflection method. *Oklahoma Geology Notes* 62:157–66.

Fay, R. O. 1988. I-35 roadcuts: Geology of Paleozoic strata in the Arbuckle Mountains of southern Oklahoma. In *South-Central Section of the Geological Society of America*, centennial field guide, vol. 4, 183–88.

Ham, W. E., et al. 1969. *Regional geology of the Arbuckle Mountains, Oklahoma.* Oklahoma Geological Survey guidebook 17.

67. Mount Mazama, Oregon

Bacon, C. R. 1983. Eruptive history of Mount Mazama and Crater Lake caldera, Cascade Range. *Journal of Volcanology and Geothermal Research* 18 (1–4):57–115.

Cranson, K. R. 2005. *Crater Lake, gem of the Cascades: The geological story of Crater Lake National Park, Oregon.* Lansing, MI: KRC Pub.

Crater Lake National Park. 1990. In *Geology of national parks*, eds. A. G. Harris, et al., 416–28. Dubuque, IA: Kendall Hunt Pub. Co.

Harris, S. L. 1988. *Fire mountains of the west: The Cascade and Mono Lake volcanoes.* Missoula, MT: Mountain Press Pub. Co.

68. Lava River Cave, Oregon

Alt, D. D., and D. W. Hyndman. 2000. *Roadside geology of Oregon.* Missoula, MT: Mountain Press Pub. Co.

Greeley, R. 1971. *Geology of selected lava tubes in the Bend area, Oregon.* Dept. of Geology and Mineral Industries bulletin 71.

Larson, C. V., ed. 1982. *An introduction to caves of the Bend area: Guidebook of the 1982 NSS Convention.* Huntsville, AL: National Speleological Society.

Larson, C., and J. Larson. 1987. *Lava River Cave.* Vancouver, WA: ABC Print and Pub.

69. Drake's Folly, Pennsylvania

Black, B. 2000. *Petrolia: The landscape of America's first oil boom.* Baltimore, MD: Johns Hopkins Univ. Press.

Giddens, P. H. 1938. *Birth of the oil industry.* New York, NY: The Macmillan Co.

Sherman, J. 2002. *Drake Well Museum and Park.* Mechanicsburg, PA: Stackpole Books.

Yergin, D. 1991. *The prize: The epic quest for oil, money, and power.* New York, NY: Simon and Schuster.

70. Hickory Run, Pennsylvania

Reiter, N. A. 2006. The Avalon Foundation. *The ringing rocks of Pennsylvania: A musical history along the Delaware.* http://www.theavalonfoundation.org/docs/rrocks.html.

Sevon, W. D. 1987. The Hickory Run boulder field, a periglacial relict, Carbon County, Pennsylvania. In *Northeastern Section of the Geological Society of America*, centennial field guide, vol. 5, 75–76.

Smith, H. T. U. 1953. The Hickory Run boulder field, Carbon County, Pennsylvania. *American Journal of Science* 251:625–42.

Van Diver, B. B. 1990. *Roadside geology of Pennsylvania.* Missoula, MT: Mountain Press Pub. Co.

71. Delaware Water Gap, Pennsylvania

Brodhead, L. W. 2010. *The Delaware Water Gap: Its scenery, its legends and early history.* Memphis, TN: General Books LLC.

Obiso, L. 2008. *Delaware Water Gap National Recreation Area: Images of America.* Charleston, SC: Arcadia Pub.

Yolton, J. S. 1980. *The Delaware River and the flow of time: Delaware Water Gap National Recreation Area.* National Park Service.

72. Beavertail Point, Rhode Island

Murray, D. P. 1988. *Rhode Island: The last billion years.* Kingston, RI: Dept. of Geology, Univ. of Rhode Island.

Quinn, A. W. 1973. *Rhode Island geology for the non-geologist.* Providence, RI: Rhode Island Dept. of Natural Resources.

Raymo, C., and M. E. Raymo. 2001. *Written in stone: A geological history of the northeastern United States.* Delmar, NY: Black Dome Press Corp.

Redfern, R. 1983. *The making of a continent.* New York, NY: Times Books.

73. Crowburg Basin, South Carolina

Bell, H., III. 1974. *Geology of the Piedmont and coastal plain near Pageland, South Carolina, and Wadesboro, North Carolina.* Carolina Geological Society guidebook for 1974 annual meeting.

Murphy, C. H. 1995. *Carolina rocks! The geology of South Carolina.* Orangeburg, SC: Sandlapper Pub. Co.

Olsen, P. E., et al. 1991. Rift basins of early Mesozoic age. In *The geology of the Carolinas*, Carolina Geological Society fiftieth anniversary volume, eds. J. W. Horton Jr. and V. A. Zullo, 142–70. Knoxville, TN: Univ. of Tennessee Press.

Stewart, K. G. 2007. *Exploring the geology of the Carolinas: A field guide to favorite places from Chimney Rock to Charleston.* Chapel Hill, NC: Univ. of North Carolina Press.

74. Mount Rushmore, South Dakota

Borglum, L. 1997. *My father's mountain: Mount Rushmore National Memorial and how it was carved.* Rapid City, SD: Fenwyn Press.

Greis, J. P. 1996. *Roadside geology of South Dakota.* Missoula, MT: Mountain Press Pub. Co.

Larner, J. 2002. *Mount Rushmore: An icon reconsidered.* New York, NY: Nation Books.

US-Parks.com. *Mount Rushmore National Memorial—geology.* http://www.us-parks.com/mount-rushmore-national-memorial/geology.html.

75. Mammoth Site, South Dakota

Greis, J. P. 1996. *Roadside geology of South Dakota.* Missoula, MT: Mountain Press Pub. Co.

Lister, A., and P. Bahn. 1994. *Mammoths*. New York, NY: Prentice Hall/Macmillan Pub. Co.

Martin, P. S. 2007. *Twilight of the mammoths: Ice Age extinctions and the rewilding of America*. Berkeley, CA: Univ. of California Press.

Nelson, L. W. 2005. *The Mammoth site: A treasure of clues to the past*. Rapid City, SD: Fenwyn Press.

76. Pinnacles Overlook, South Dakota

Badlands National Park. 1990. In *Geology of national parks*, eds. A. G. Harris, et al., 110–23. Dubuque, IA: Kendall Hunt Pub. Co.

Durant, M., and M. Harwood. 1988. *This curious country: Badlands National Park*. Interior, SD: Badlands Natural History Assoc.

Martin, J. E. 1987. The White River badlands of South Dakota. In *Rocky Mountain Section of the Geological Society of America*, centennial field guide, vol. 2, 233–36.

Zarki, J. W. 2008. *Badlands: The story behind the scenery*. Las Vegas, NV: KC Publications Inc.

77. Reelfoot Scarp, Tennessee

Bagnall, N. H. 1996. *On shaky ground: The New Madrid earthquakes of 1811–1812*. Columbia, MO: Univ. of Missouri Press.

Earthquakes in Missouri. 1988. Rolla, MO: Missouri Dept. of Natural Resources, Division of Geology and Land Survey.

Johnston, A. C., and E. S. Schweig. 1996. The enigma of the New Madrid earthquakes of 1811–1812. *Annual Review of Earth and Planetary Science* 24:339–84.

Stewart, D., and R. Knox. 1996. *The Earthquake that never went away: The shaking stopped in 1812, but the impact goes on*. Marble Hill, MO: Gutenberg-Richter Publications.

78. Enchanted Rock, Texas

Barnes, V. E. 1988. The Precambrian of Central Texas. In *South-Central Section of the Geological Society of America*, centennial field guide, vol. 4, 361–68.

Hutchinson, R. M. 1988. Enchanted Rock dome, Llano and Gillespie counties, Texas. In *South-Central Section of the Geological Society of America*, centennial field guide, vol. 4, 369–71.

Spearing, D. 1991. *Roadside geology of Texas*. Missoula, MT: Mountain Press Pub. Co.

Walker, N. 1992. Middle Proterozoic geologic evaluation of the Llano Uplift, Texas: Evidence from U-Pb zircon geochronometry. In *Geological Society of America Bulletin* 104 (4):494–504.

79. Capitan Reef, Texas

Bebout, D. G., and C. Kerans. 1993. *Guide to the Permian Reef Geology Trail, McKittrick Canyon, Guadalupe National Park, West Texas*. Austin, TX: Texas Bureau of Economic Geology.

Budd, A. F. 1990. Guadalupe Mountains National Park. In *Geology of national parks*, eds. A. G. Harris, et al., 177–85. Dubuque, IA: Kendall Hunt Pub. Co.

Jagnow, D. H., and R. R. Jagnow. 1992. *Stories from stones: The geology of the Guadalupe Mountains*. Carlsbad, NM: Carlsbad Caverns Guadalupe Mountains Assoc.

Murphy, D. 1984. *The Guadalupes: Guadalupe National Park*. Carlsbad, NM: Carlsbad Caverns Guadalupe Mountains Assoc.

80. Paluxy River Tracks, Texas

Bird, R. T. 1985. *Bones for Barnum Brown: Adventures of a dinosaur hunter*. Fort Worth, TX: Texas Christian Univ. Press.

Farlow, J. O. 1993. *The dinosaurs of Dinosaur Valley State Park*. Austin, TX: Texas Parks and Wildlife Dept.

Gregory, P. S. 2010. *The Princeton field guide to dinosaurs*. Princeton, NJ: Princeton Univ. Press.

Thulborn, T. 1990 *Dinosaur tracks*. London, UK: Chapman and Hall.

81. Upheaval Dome, Utah

Chronic, H. 1990. *Roadside geology of Utah*. Missoula, MT: Mountain Press Pub. Co.

Huntoon, P. W. 2000. Upheaval Dome, Canyonlands, Utah: Strain indicators that reveal an impact origin. In *Geology of Utah's parks and monuments*, Utah Geological Assoc. publication 28, eds. D. A. Sprinkel, et al., 1–10.

Jackson, M. P. A., et al. 1998. Structure and evolution of Upheaval Dome: A pinched-off salt diapir. *Geological Society of America Bulletin* 110 (12):1547–73.

Kriens, B. J., et al. 1999. Geology of the Upheaval Dome impact structure, Southeast Utah. *Journal of Geophysical Research* 104 (E8):18867–87.

82. Checkerboard Mesa, Utah

Chan, M. A., and A. W. Archer. 2000. Cyclic eolian stratification on the Jurassic Navajo sandstone, Zion National Park: Periodicities and implications for paleoclimate. In *Geology of Utah's parks and monuments*, Utah Geological Assoc. publication 28, eds. D. A. Sprinkel, et al., 606–17.

Hamilton, W. L., et al. 1984. *The sculpturing of Zion: Guide to the geology of Zion National Park*. Springdale, UT: Zion Natural History Assoc.

Orndorff, R. L., et al. 2006. *Geology underfoot in southern Utah*. Missoula, MT: Mountain Press Pub. Co.

Parrish, J. T. 1993. Climate of the supercontinent Pangaea. *Journal of Geology* 101 (2):215–33.

83. San Juan Goosenecks, Utah

Baars, D., and G. Stevenson. 1986. *San Juan canyons: A river runner's guide and natural history of San Juan canyons*. Cañon Publications Ltd.

Chronic, H. 1990. *Roadside geology of Utah*. Missoula, MT: Mountain Press Pub. Co.

Orndorff, R. L., et al. 2006. *Geology underfoot in southern Utah*. Missoula, MT: Mountain Press Pub. Co.

Stevenson, G. M. 2000. Geology of Goosenecks State Park, Utah. In *Geology of Utah's parks and monuments*, Utah Geological Assoc. publication 28, eds. D. A. Sprinkel, et al., 433–47.

84. Salina Canyon Unconformity, Utah

Chronic, H. 1990. *Roadside geology of Utah*. Missoula, MT: Mountain Press Pub. Co.

Fillmore, R. 2011. *Geological evolution of the Colorado Plateau of eastern Utah and western Colorado*. Salt Lake City, UT: Univ. of Utah Press.

Hintze, L. F., and B. J. Kowallis. 2009. *Geologic history of Utah: A field guide to Utah's rocks*. Provo, UT: Dept. of Geology, Brigham Young Univ.

Lawton, T. F., and G. C. Willis. 1987. The geology of Salina Canyon, Utah. In *Rocky Mountain Section of the Geological Society of America*, centennial field guide, vol. 2, 265–68.

85. Bingham Stock, Utah

Chronic, H. 1990. *Roadside geology of Utah*. Missoula, MT: Mountain Press Pub. Co.

James, L. P. 1978. The Bingham copper deposits, Utah, as an exploration target: History and pre-excavation geology. *Economic Geology* 73 (7):1218–27.

Lanier, G., et al. 1978. General geology of the Bingham Mine, Bingham Canyon, Utah. *Economic Geology* 73 (7):1228–41.

Shelton, J. S. 1966. *Geology illustrated*. San Francisco, CA, and London, UK: W. H. Freeman and Co.

86. Whipstock Hill, Vermont

Doolan, B. L. 1996. The geology of Vermont. *Rocks and Minerals* 71 (4):218–25.

Raymo, C., and M. E. Raymo. 2001. *Written in stone: A geological history of the northeastern United States*. Delmar, NY: Black Dome Press Corp.

St. John, J. *Whipstock Hill outcrop*. http://www1.newark.ohio-state.edu/Professional/OSU/Faculty/jstjohn/Vermont-Geology/Whipstock-Hill.htm.

Van Diver, B. B. 1987. *Roadside geology of Vermont and New Hampshire*. Missoula, MT: Mountain Press Pub. Co.

87. Great Falls, Virginia

Dietrich, R. V. 1990. *Geology and Virginia*. Charlottesville, VA: Virginia Division of Mineral Resources.

Frye, K. 1986. *Roadside geology of Virginia*. Missoula, MT: Mountain Press Pub. Co.

Reed, J. C., et al. 1980. *The river and the rocks: The geologic story of Great Falls and the Potomac River Gorge*. U.S. Geological Survey bulletin 1471.

Virginiaplaces.org. *The fall line*. http://www.virginiaplaces.org/regions/fallshape.html.

88. Natural Bridge, Virginia

Catlin, D. T. 1984. *A naturalist's Blue Ridge Parkway*. Knoxville, TN: Univ. of Tennessee Press.

Frye, K. 1990. *Roadside geology of Virginia*. Missoula, MT: Mountain Press Pub. Co.

Spencer, E. W. 1968. *Geology of the Natural Bridge, Sugarloaf Mountain, Buchanan, and Arnold Valley quadrangles, Virginia*. Virginia Division of Mineral Resources, report of investigations 13.

———. 1985. *Guidebook to the Natural Bridge and Natural Bridge Caverns*. Lexington, VA: Poorhouse Mountain Studios.

89. Millbrig Ashfall, Virginia

Frye, K. 1990. *Roadside geology of Virginia*. Missoula, MT: Mountain Press Pub. Co.

Huff, W. D., et al. 1992. Gigantic Ordovician volcanic ash fall in North America and Europe: Biological, tectonomagmatic, and event-stratigraphic significance. *Geology* 20 (10):875–78.

Huffman, G. G. 1945. Middle Ordovician limestones from Lee County, Virginia, to central Kentucky. *Journal of Geology* 53 (3): 145–74.

Kolata, D. R., et al. 1996. *Ordovician K-bentonites of eastern North America*. Geological Society of America special paper 313.

90. Catoctin Greenstone, Virginia

Badger, R. L., and A. K. Sinha. 1988. Age and Sr isotopic signature of the Catoctin volcanic province: Implications for subcrustal mantle evolution. *Geology* 16 (8):692–95.

Bailey, C. M. Mind the gap! Where is Afton Mountain? In *The geology of Virginia*. Dept. of Geology, College of William and Mary. http://web.wm.edu/geology/virginia/field_loc/rfgweb.pdf.

Bentley, C. 2008. *An overview of Shenandoah National Park's geologic story*. Northern Virginia Community College. http://www.nvcc.edu/home/cbentley/shenandoah/index.htm.

Reed, J. C., Jr. 1955. Catoctin Formation near Luray, Virginia. *Geological Society of America Bulletin* 66 (7):871–96.

91. Mount St. Helens, Washington

Corcoran, T. 2006. *Mount St. Helens: The story behind the scenery*. Las Vegas, NV: KC Publications Inc.

Doukas, M. P., and D. A. Swanson. 1987. Mount St. Helens, Washington, with emphasis on 1980–85 eruptive activity as viewed from Windy Ridge. In *Cordilleran Section of the Geological Society of America*, centennial field guide, vol. 1, 333–38.

Sherrod, D. R., et al., eds. 2008. *A volcano rekindled: The renewed eruption of Mount St. Helens, 2004–2006*. U.S. Geological Survey professional paper 1750.

Williams, C. 1980. *Mount St. Helens: A changing landscape*. Portland, OR: Graphic Arts Center Pub. Co.

92. Dry Falls, Washington

Alt, D. D. 2001. *Glacial Lake Missoula and its humongous floods*. Missoula, MT: Mountain Press Pub. Co.

Alt, D. D., and D. W. Hyndman. 1984. *Roadside geology of Washington*. Missoula, MT: Mountain Press Pub. Co.

Bretz, J. H. 1928. The channeled scabland of eastern Washington. *Geographical Review* 18 (3):446–77.

Weis, P., and W. L. Newman. 1999. *The channeled scablands of eastern Washington: The geologic story of the Spokane flood*. Cheney, WA: Eastern Washington Univ. Press.

93. Seneca Rocks, West Virginia

Clauson-Wicker, S. 2009. *West Virginia off the beaten path: A guide to unique places*. Guilford, CT: GPP Travel.

Fichter, L. S., and S. J. Baedke. The late Paleozoic Alleghanian orogeny. In *The geological evolution of Virginia and the mid-Atlantic region*. http://csmres.jmu.edu/geollab/vageol/vahist/K-LatPal.html.

Hörst, E. J. 2001. *Rock climbing Virginia, West Virginia, and Maryland*. Helena, MT: Falcon Press.

Perry, W. J., Jr. 1978. *The Wills Mountain anticline: A study in complex folding and faulting in eastern West Virginia*. West Virginia Geological and Economic Survey report of investigation RI 32.

94. Roche-A-Cri Mound, Wisconsin

Clayton, L., and J. W. Attig. 1989. *Glacial Lake Wisconsin*. Geological Society of America memoir 173.

Crowns, B. 1976. *Wisconsin through 5 billion years of change*. Wisconsin Rapids, WI: Wisconsin Earth Science Center.

Dott, R. H., Jr., and J. W. Attig. 2004. *Roadside geology of Wisconsin*. Missoula, MT: Mountain Press Pub. Co.

Mickelson, D. M., et al. 2011. *Geology of the Ice Age National Scenic Trail*. Madison, WI: Univ. of Wisconsin Press.

95. Van Hise Rock, Wisconsin

Attig, J. W., et al. 1990. *The ice age geology of Devils Lake State Park*. Wisconsin Geological and Natural History Survey educational series 35.

Dalziel, I. W. D., and R. H. Dott, Jr. 1970. *Geology of the Baraboo District, Wisconsin*. Wisconsin Geological and Natural History Survey information circular 14.

Dott, R. H., Jr., and J. W. Attig. 2004. *Roadside geology of Wisconsin*. Missoula, MT: Mountain Press Pub. Co.

Hubbard, E. 2010. *Dr. Charles R. Van Hise*. Whitefish, MT: Kessinger Pub. LLC.

96. Amnicon Falls, Wisconsin

Allen, D. J., et al. 1997. Integrated geophysical modeling of the North American Midcontinent Rift System: New interpretations for western Lake Superior, northwestern Wisconsin, and eastern Minnesota. In *Middle Proterozoic to Cambrian rifting, central North America*, Geological Society of America special paper 312, eds. R. W. Ojakangas, et al., 47–72.

Dickas, A. B. 1984. Midcontinent Rift System: Precambrian hydrocarbon target. *Oil and Gas Journal* 82:151–59.

———. 1986. Comparative Precambrian stratigraphy and structure along the Midcontinent Rift. *American Association of Petroleum Geologists Bulletin* 70:225–38.

Dickas, A. B., and M. G. Mudrey Jr. 1997. Segmented structure of the middle Proterozoic Midcontinent Rift System, North America. In *Middle Proterozoic to Cambrian rifting, central North America*, Geological Society of America special paper 312, eds. R. W. Ojakangas, et al., 37–46.

97. Green River, Wyoming

Blackstone, D. L., Jr. 1988. *Traveler's guide to the geology of Wyoming*. Geological Survey of Wyoming bulletin 67.

Grande, L. 1984. *Paleontology of the Green River Formation, with a review of the fish fauna*. Geological Survey of Wyoming bulletin 63.

Lageson, D. R., and D. Spearing. 1988. *Roadside geology of Wyoming*. Missoula, MT: Mountain Press Pub. Co.

Lavelle, M. 2006. The new oil rush: High tech is transforming the market. *U.S. News and World Report* 140 (15):42–49.

98. Devils Tower, Wyoming

Blackstone, D. L., Jr. 1988. *Traveler's guide to the geology of Wyoming*. Geological Survey of Wyoming bulletin 67.

Karner, F. R., and D. L. Halvorson. 1987. The Devils Tower, Bear Lodge Mountains, Cenozoic igneous complex, northeastern Wyoming. In *Rocky Mountain Section of the Geological Society of America*, centennial field guide, vol. 2, 161–64.

Lageson, D. R., and D. Spearing. 1988. *Roadside geology of Wyoming.* Missoula, MT: Mountain Press Pub. Co.

Robinson, C. S. 1956. *Geology of Devils Tower National Monument, Wyoming.* U.S. Geological Survey bulletin 1021-I.

99. Fossil Butte, Wyoming

Ambrose, P. D. 1996. *Fossil Butte National Monument: Along the shores of time.* Vernal, UT: Dinosaur Nature Assoc.

Caggiano, T. *The Green River Formation.* http://fossilnews.com/2000/grnrv/grnrv.html.

Grande, L. 1984. *Paleontology of the Green River Formation, with a review of the fish fauna.* Geological Survey of Wyoming bulletin 63.

McGrew, P. O., and M. Casilliano. 1975. *The geological history of Fossil Butte National Monument and Fossil Basin.* National Park Service occasional paper 3.

100. Steamboat Geyser, Wyoming

Brantley, S. R. 2004. *Tracking changes in Yellowstone's restless volcanic system.* U.S. Geological Survey fact sheet 100-03.

Bryan, T. S. 2005. *Geysers: What they are and how they work.* Missoula, MT: Mountain Press Pub. Co.

Foley, D. 2006. *Yellowstone's geysers: The story behind the scenery.* Las Vegas, NV: KC Publications Inc.

Lowenstern, J. B., et al. 2005. *Steam explosions, earthquakes, and volcanic eruptions—what's in Yellowstone's future.* U.S. Geological Survey fact sheet 2005-3024.

101. Specimen Ridge, Wyoming

Beyer, A. F. 1954. Some petrified wood from Specimen Ridge area of Yellowstone National Park. *American Midland Naturalist* 51 (2):553–76.

Knowlton, F. H. *The fossil forests of the Yellowstone National Park.* U.S. Geological Survey. http://www.nps.gov/history/history/online_books/yell/knowlton/sec1.htm.

Lageson, D. R., and D. Spearing. 1988. *Roadside geology of Wyoming.* Missoula, MT: Mountain Press Pub. Co.

Tuttle, S. D. 1990. Yellowstone National Park. In *Geology of national parks*, eds. A. G. Harris, et al., 480–500. Dubuque, IA: Kendall Hunt Pub. Co.

Index

abrasion, 158
Absaroka Mountains, 220
Acadia National Park, 80
Acadian Mountains, 72, 73, 80
Acadian orogeny, 14, 162
acanthodians, 14
Acasta gneiss, 111
Acrocanthosaurus, 178, 179
Africa, 5, 6, 8
Agassiz, Louis, 148
age dating: absolute, 4–5; relative, 1–3
Age of Dinosaurs, 42
Age of Fishes, 14, 72
Age of Grasses, 17
Age of Ice, 17
Age of Mammals, 16
Age of Man, 17
Age of Oil, 38
Age of Petroleum, 212
Age of Prokaryotes, 12
Age of Reptiles, 16
Agricola, Georgius, 1
Alabama, 20
Alaska, 22
Aleutian Islands, 152
Aleutian Trench, 56
Algoman orogeny, 96, 136, 137
Alleghanian orogeny, 14, 136, 137, 162
Allegheny Front, 204
allosaurus, 16
Alps, 10, 15
Altyn limestone, 104
American Falls, 134, 135
Amnicon Falls, 96, 97
Amnicon Falls State Park, 210
amphibians, 14, 72
amygdaloid copper, 89
ancestral Rockies, 42. *See also* Rocky Mountains

Andes Mountains, 9, 17
angular unconformity, 120, 121, 136, 137, 186, 187
Antarctica, 6, 13, 15, 132, 142
Antelope Canyon, 24, 25
Anthropocene, 18
anticlines, 94, 95, 150
Antler terrane, 14
apatosaurus, 44
Apennines, 17
Appalachian Basin, 72, 76, 77
Appalachian Mountains, 14, 15, 50, 54, 124, 136, 162, 204
Appalachian Plateau, 204
Appekunny Formation, 104
applied geology, 1
Arbuckle Basin, 150
Arbuckle Mountains, 150
Archean eon, 12
Arctica, 110
Argentinosaurus, 16
Arikaree Formation, 142, 143
Arizona, 24, 26, 28
Arkansas, 14, 30
Ashfall Fossil Beds State Historical Park, 112, 113
Ash Hollow Formation, 112
asphalt, 38
astroblemes, 20, 26
Atlantic Coastal Plain, 192, 193
Atlantic Ocean, 9, 124, 164
Atlas Range, 10, 17
Austin Glen Formation, 136, 137
Australia, 176
Avalonia, 80, 162
Avalon terrane, 162, 163
Avery Island, 78, 79

Bacon, Sir Francis, 5

badlands, 170, 171
Badlands National Park, 170, 171
Baikal, Lake, 108
Baltica, 13
banded iron formation, 12, 96, 97
Baraboo District, 208
Baraboo quartzite, 208
basalt, 1, 36, 37, 54, 94, 95
Bascom Formation, 190, 191
Basin and Range Province, 58, 59
basin formation, 64, 65
Bayfront Park, 82, 83
Beartooth Highway, 110
Beartooth Plateau, 110
Beavertail Point, 162, 163
Belt Supergroup, 104, 105
Bend flow, 154, 155
Benton Formation, 42
Bering Land Bridge, 168
Berkeley Pit, 108, 109
Big Island, 56, 60
Big Pinnacle, 140, 141
Big Rock, 146, 147
Big Room, 128, 129
Big Sandy Gas Field, 72, 73
Bingham Canyon, 188
Bingham Canyon Mine, 188
Bingham Stock, 188, 189
Black Beauty diamond, 31
Black Hills, 166, 167
black smokers, 11
blue-green algae, 12, 212
Blue Ridge Mountains, 17
Blue Ridge Province, 204
Boltwood, Bertram, 5
Bonneterre Formation, 102, 103
bony fish, 14, 70
booming sands, 118
Borah Peak, 58, 59

boudinage, 198
boudins, 199
Boulder Batholith, 108, 109
boulder fields, 158
Boulder Flatirons, 42, 43
brachiopods, 144, 145, 151
Brachiosaurus, 48
Brandywine Blue Gneiss, 50
Brandywine Creek State Park, 50
Breathitt Formation, 76, 77
Bretz, J. Harlen, 202
brontosaurus, 44
Brownies Beach, 82
Bruhathkayosaurus, 16
Brule Formation, 114, 170, 171
bryozoans, 144, 150, 151
Butte Pluton, 108, 109
Butterfield Overland Stage Route, 71

calcite, 128
calcium carbonate, 74
caldera, 152
California, 32, 34, 36, 38, 40, 130
Calvert Cliffs, 82, 83
Calvert Formation, 82
Cambrian explosion, 13, 66, 102, 136
Cambrian period, 13
Canyon Diablo Meteor, 26, 27
Canyonlands National Park, 24, 180, 181
canyons, slot, 24
Capitan Reef, 176, 177
Capitol Reef National Park, 24
carbonate rock. *See* limestone
carbonic acid, 128
Carboniferous period, 14
Cardiff Giant, 68
Carlsbad Caverns, 128, 129
Cascade Range, 10, 130, 200, 201
castellated mounds, 206, 207
Castile Formation, 132
cataracts, 134
catastrophism, 2, 90
Catlin, George, 92
catlinite, 92, 93
Catoctin greenstone, 198, 199
Catskill Formation, 158

"cat's paws," 54
caverns, 127, 128
caves, 74
Cenozoic era, 13, 16
Chadron Formation, 142, 170, 171
chalk, 70
Chalk Pyramids, 70
Champsosaurus, 142
Channeled Scablands, 202
Checkerboard Mesa, 182, 183
chert, 96
Chesapeake Bay, 17, 82
Chicxulub Crater, 16
Chief Mountain, 104, 105
Chihuahuan Desert, 176
Chimney Rock, 114
Chinese Theatre, 129
Choptank Formation, 82
cinder cones, 60, 62
claystone, 92
cleavage, 208
climate change, 15, 17
Close Encounters of the Third Kind, 214
Clovis culture, 168
coal, 14, 76, 77
Coast Range, 16
coesite, 26
Colorado, 42, 44, 46
Colorado Mineral Belt, 42
Colorado Plateau, 26, 29, 180, 181, 185
Colton Formation, 186
columnar basalt, 36, 37
columnar fractures, 94
columnar structure, 214
Conanicut Island, 162
concretions, 150
conglomerate, 1, 52
Connecticut, 48, 49
continental drift, 7–8
convection cells, 9, 10, 122
convergent plate boundary, 9, 10, 11
copper, 86, 87, 88, 89, 188
coral reef, 66
Coral Sea, 16
core, Earth's, 9, 10, 11
Courthouse Rock, 114

Crater Lake, 152, 153
Crater Lake National Park, 152
Crater of Diamonds State Park, 30
craters, 20, 21, 26, 27
Craters of the Moon National Monument and Preserve, 62
craton, 210, 211
Cretaceous period, 16
crinoids, 144, 151
crossbedding, 24
crossbeds, 90, 91, 115, 182, 183
crosscutting relationships, principle of, 2, 3
crossopterygian, 72
Crowburg Basin, 164
Crow Creek marlstone, 116, 117
crust, 1, 10, 11, 12, 52
cryptovolcanic explosive structure, 20
Crystal Ice Cave, 62

Dakota Group, 44, 45
Daniels, Edward, 208
dark matter, 96
dawn-horse, 17
Death Valley, 34
Death Valley National Park, 34
Deccan lava flows, 16, 94
DeChelly Formation, 28
Delaware, 50
Delaware Basin, 132, 133, 176, 177
Delaware Sea, 128
Delaware Water Gap, 160, 161
Denver, 44, 45
deserts, 126
Devil's Millhopper Geological State Park, 52, 53
Devils Tower, 214, 215
Devonian period, 14
Diamond Head, 60
diamonds, 30, 31
diatreme, 130, 131
dikes, 54, 80, 81, 94, 103, 130, 131
Dilophosaurus, 48, 49
dinosaurs, 15, 16, 262, 263
Dinosaur State Park, 48, 49
Dinosaur Valley State Park, 178, 179
dire wolf, 38

disconformity, 120, 121
divergent plate boundary, 9, 10, 11
dolomite, 34
Drake Well, 156, 157
Drake Well Museum, 156
Driftless Area, 206, 207
Dry Falls, 202, 203
Dry Falls Lake, 203
Dutch Island Harbor Formation, 163

Eads Bridge, 100
Early Paleogene Thermal Maximum, 17
Earth: age of, 1–5; formation of, 11; history of, 2, 3, 11–18; uniqueness of, 94
Earthquake Lake, 107
Earthquake Lake Visitor Center, 106
earthquakes, 10, 11. *See also* Borah Peak; Purgatory Chasm; Reelfoot Scarp; Wallace Creek
East African Rift, 9
East African Rift System, 17
Eastern Kentucky Coal Field, 76, 77
East Mitten Butte, 29
Edwards limestone, 174
Edwards Plateau, 174
ejecta, 26
El Capitan (CA), 40, 41
El Capitan (TX), 177
Elephant Rocks, 100, 101
Elephant Rocks State Park, 100, 101
El Niño, 66
Enchanted Rock, 174, 175
Enchanted Rock State Natural Area, 174
eon, 2
epoch, 2
era, 2
Erie, Lake, 134
erosion: and El Capitan, 40; flood, 202, 203; headwater, 160, 161; inverted topography, 142, 143; karst, 52; mass movement, 106, 107; principal forces of, 28; stream, 86, 87, 160, 161, 184, 185; wind, 34, 70, 71. *See also* exfoliation

Everest, Mount, 10
evolutionary opportunism, 14
exfoliation, 54, 55, 174, 175
Exit Glacier, 22, 23
exotic terranes, 162, 163
extinction events, 15, 16, 28, 48
extrusive igneous rock, 1

fall line, 20, 192
faults, 32, 58, 104, 105, 107
faunal succession, principle of, 82
Fermi National Accelerator Laboratory, 96
fin-and-crag topography, 204, 205
fish within a fish fossil, 70, 71
Fitzgerald Park, 90, 91
Five Islands, 78
Flatirons, 42
floods, 24
Florida, 52
Florissant Formation, 46, 47
Florissant Fossil Beds, 46, 47
Florissant Fossil Beds National Monument, 46
flow banding, 54
flow lines, 81
flowstone, 74, 128
Flume Gorge, 122, 123
Foerstia, 72
foraminifera, 70
Forest Hill Formation, 98
forests, 14, 138
Fort Dodge Formation, 68, 69
Fort Dodge gypsum, 68, 69
Fossil Butte, 216, 217
Fossil Butte National Monument, 216
fossils: casts, 138, 139; dinosaur tracks, 48, 49, 178, 179; fish within a fish, 70, 71; lungfish, 72; mammal, 112, 113; mammoth, 168; marine reptile, 70; oldest-known tree, 138, 139; Ordovician-age, 144; Paleogene-age, 216; petrified forests, 98, 99, 220, 221; Pleistocene, 38, 39; redwood, 221; reef, 66, 67, 176, 177;

rhinoceroses, 113; sand dunes, 182, 183; seaweed, 72; shark teeth, 82
Fountain Formation, 42
Four Corners region, 130
Franconia Notch State Park, 122
Franklin, Benjamin, 5
Frenchman Mountain, 120, 121
Front Range, 42, 43
frost wedging, 84, 166
Fuji, Mount, 10

gabbro, 1
gastropods, 151
gelifluction, 158
genesis rocks, 110
geologic time, 2, 4, 5, 11. *See also individual eons; epochs; eras; periods*
geology, principles of, 1, 2, 3
Georgia, 54
geothermal energy, 60
Gilboa Forest, 138, 139
Gillicus, 70
glacial episodes, 12, 14, 17, 18, 22
glacial floods, 202, 203
glacial grooves, 148, 149
Glacial Grooves State Memorial, 148
Glacial Lake Missoula, 202, 203
glacial polish, 36, 37
glacial striations, 37, 84, 146, 147, 148, 149
glaciation, 146, 147. *See also* glacial episodes
Glacier National Park, 22, 104
glaciers, 22, 23
global warming, 18, 22, 132, 176, 178. *See also* climate change
gneiss, 1, 5, 50, 110
Golden Valley Formation, 142
Gondwanaland, 6, 13, 14
graben, 124, 210, 211
granite, 1, 40, 54, 122, 123
Granite State, 14, 122
Grassy Mountain ignimbrite, 102
Great Appalachian Valley, 160, 194
Great Barrier Reef, 176
Great Falls, 192, 193
Great Falls Park, 192, 193

245

Great Lakes, 18, 134
Great Plains, 92, 114
Great Rift, 62, 63
Great Unconformity, 120, 121
Greenland, 12, 16
Green Mountain (CO), 43
Green Mountains (VT), 190
Green River Formation, 216, 217
Green River shale, 212, 213
greenstone, 198, 199
Grenvillia, 9, 190, 210, 211
Guadalupe Mountains National Park, 176
Guadalupe Peak, 176, 177
Guffey volcanic complex, 46, 47
Gulf of Mexico, 15, 70, 78
Gunsight Notch, 204, 205
gypsum, 68, 69, 126

Hadean eon, 11–12
Half Dome, 41
half graben, 48, 49
Hall, Sir James, 186
Hanging Rock klint, 66, 67
Harding Ice Field, 22
Harney Peak, 166
Hartford Basin, 48, 49, 164
Hawaii, 11, 56, 57, 60
Hawaiian Island–Emperor Seamount chain, 11, 56
Hawaiian Islands, 56
headwater erosion, 160, 161
Hebgen Lake, 106
Hickory Run boulder field, 158, 159
High Cascades, 152, 153
Himalaya Range, 10, 17, 94
Holmes, Arthur, 5
Holocene epoch, 18
Honaker Trail Formation, 184
Hood, Mount, 154
Hooke, Robert, 1
horn corals, 144, 145
horses, 17
Horseshoe Falls, 134
horst, 210, 211
hot spots, 11, 56, 57, 218, 219
hot springs, 11

Hubbert, M. King, 212
Hudson Bay, 5
Hueston Woods State Park, 144, 145
Humboldt, Alexander von, 5
Huron, Lake, 134
Hutton, James, 2, 186, 187
hydrothermal fluids, 188

Iapetus Ocean, 13, 50, 51, 190
ice age. *See* glacial episodes
Iceland, 130
Idaho, 58, 60, 62
igneous rocks, 1, 52
ignimbrite, 102
Iguanodon, 178
Illinois, 64
Illinois Basin, 64, 65
inclusions, principle of, 2
Indiana, 66
insects, 17, 46
inselberg, 140, 175
Interstate Highway System, 44
intrusions, igneous, 1, 30, 31, 36, 37, 40
inverted topography, 142, 143
Iowa, 68
iridium, 16, 20
iron ore, 97
island arc, 9
Isotelus, 144

Jail Rock, 114
Jamestown Formation, 163
Japan, 152
Jefferson, Thomas, 194
Johnston Ridge Observatory, 200
Jurassic period, 15–16

Kansas, 12, 70
karst, 52, 194
Kaskaskia Sea, 14, 72, 73
Kelleys Island, 148, 149
Kenai Fjords National Park, 22
Kenorland, 110
Kentucky, 72, 74, 76
kerogen, 212, 213
Keweenaw Peninsula, 86, 88, 89

Kilauea Volcano, 56, 57
Kings Bowl, 63
Kittatinny Mountain, 160, 161
klippe, 104
Knightia, 217
Krakatoa, 196

Labrador, 12
laccolith, 214, 215
lacewing, 47
lagerstätten, 216
Lake Vermilion Gold Rush, 96
lamproite, 30, 31
landslides, 106, 107
Laramide orogeny, 17, 28, 128
lateral continuity, principle of, 2, 3, 114, 115
Laurasia, 14
Laurentia, 12, 13, 14, 50
lava, 88, 154, 155
lavacicles, 154
Lava River Cave, 154, 155
lava tubes, 154
Lewis Overthrust, 104
life-forms, 12, 13
limestone, 1, 52, 53, 74, 75, 194, 195
Lincoln Brick Park, 90, 91
Little Ice Age, 18
Llano Uplift, 174
lobe-fin, 14
Lockatong Formation, 124, 125
Lo'ihi, 56, 57
Los Angeles, 32
Louann Salt, 78
Lucero, Lake, 126
lungfish, 72
Lykins Formation, 42
Lyons Formation, 42
Lytle Formation, 44

maar, 130, 131
Madison Range, 106
Madison Slide, 106, 107
magma, 12. *See also* intrusions, igneous
Mahogany Ridge zone, 212
Maine, 80